In *Robert Boyle and the Limits of Reason,* Jan W. Wojcik explores the theological context within which Boyle developed his views on reason's limits. Wojcik shows how Boyle's three categories of "things above reason" – the incomprehensible, the inexplicable, and the unsociable – were reflected in his conception of the goals and methods of natural philosophy.

Throughout the book, Wojcik emphasizes Boyle's remarkably unified worldview in which truths in chemistry, physics, and theology were but different aspects of one unified body of knowledge. She concludes with an analysis of the presupposition on which Boyle's views on the limits of reason rested: that when God created intelligent beings, he deliberately chose to limit their understanding, reserving a complete understanding for the afterlife.

Robert Boyle and the Limits of Reason

Robert Boyle and the Limits of Reason

JAN W. WOJCIK

AUBURN UNIVERSITY

CAMBRIDGE
UNIVERSITY PRESS

PUBLISHED BY THE PRESS SYNDICATE OF THE UNIVERSITY OF CAMBRIDGE
The Pitt Building, Trumpington Street, Cambridge, United Kingdom

CAMBRIDGE UNIVERSITY PRESS
The Edinburgh Building, Cambridge CB2 2RU, UK
40 West 20th Street, New York NY 10011–4211, USA
477 Williamstown Road, Port Melbourne, VIC 3207, Australia
Ruiz de Alarcón 13, 28014 Madrid, Spain
Dock House, The Waterfront, Cape Town 8001, South Africa

http://www.cambridge.org

First published 1997
First paperback edition 2002

Typeface Sabon.

A catalogue record for this book is available from the British Library

Library of Congress Cataloguing in Publication data
Wojcik, Jan W.
Robert Boyle and the limits of reason / Jan W. Wojcik.
p.   cm.
Includes bibliographical references and index.
ISBN 0 521 56029 2 (hardcover)
1. Boyle, Robert, 1627–1691 – Contributions in doctrine of the
limits of human reason.   2. Reason – History – 17th century.
3. Natural theology – History of doctrines – 17th century.
4. Philosophical theology – History – 17th century.   5. Man (Christian
theology) – History of doctrines – 17th century.   I. Title.
B1201.B44W64   1997
192–dc20   96-35166   CIP

ISBN 0 521 56029 2 hardback
ISBN 0 521 52522 5 paperback

For Jim Force,

whose unwavering belief in my work
has opened many doors

Contents

Preface   page ix
Acknowledgments   xv

Introduction   1
  Robert Boyle as Lay Theologian   1
  Boyle's Early Responses to Religious Controversies   10
  Liberty of Conscience   14
  "Love of Truth" and "Love of Peace"   19

Part I   The Theological Context

1 Things above Reason: Medieval Context and Concepts   27
  Irenaeus and Clement of Alexandria   28
  Thomas Aquinas   29
  Double-truth and the Law of Noncontradiction   31
  Lorenzo Valla   35
  Two Approaches Summarized   36
  Anglicans and Puritans   37

2 The Threat of Socinianism   42
  The Protestant Background   42
  Early Socinianism   43
  The "Englishing" of Socinianism   47
  Boyle's Response to Socinianism (c. 1652)   55
  Other Responses to Socinianism   59
  Conclusions   73

3 Predestination Controversies   76
  Arminians versus Calvinists   77
  Doctrinal Issues   79
  Boyle's *Seraphic Love*   82
  Howe's *Reconcileableness* and Hammond's *Pacifick Discourse*   85

4 Theology and the Limits of Reason   95
  Style of the Scriptures   95

Reconcileableness of Reason and Religion   97
Things above Reason   100
The Charge of Enthusiasm and *Advices*   108
Conclusions   113

Part II The Context of Natural Philosophy

5 **Philosophies of Nature and their Theological Implications**   121
The Aristotelians   123
The Cambridge Platonists   126
The "Chymical" Tradition   129

6 **Sources of Knowledge**   137
Scriptural Revelation   137
Personal Revelation   141
Abstract Reason and Innate Ideas   144
Sensory Perception   146

7 **The Limits of Reason and Knowledge of Nature**   151
The Incomprehensible, the Inexplicable, and the Unsociable   152
The Task of the Natural Philosopher   161
Evaluation of Alternative Theories of Matter   168
The Question of the Falsity of Rejected Hypotheses   175
The Question of the Truth of the Corpuscular Hypothesis   179
Advantages of the Corpuscular Hypothesis   180
Some Things not Explicable By any Means   184
The Question of Progress in Natural Philosophy   186

8 **Boyle's Voluntarism and the Limits of Reason**   189
The Seventeenth-Century Background   190
Specific Aspects of Boyle's Voluntarism   200
God's Will and Human Reason   206
The Christian Virtuoso's Final Reward   210

Conclusion   212

Bibliography   220
Index   239

Preface

Being considered one of the "great" thinkers of the past has all too often resulted in a highly selective presentation of an individual's thought. In what has come to be known as a "Whiggish" interpretation of history, aspects of an individual's thought that are judged to have contributed to some perceived line of progress from the past to the present have been emphasized, while those that might be considered "embarrassing" from a twentieth-century point of view, such as a belief in witchcraft or alchemy, have been glossed over as vestiges of the past that unfortunately marred the otherwise progressive thought of the individual. One of the most encouraging aspects of recent Boyle studies is that Boyle scholars today are attempting to see his thought as a unified whole rather than selecting only those aspects that might be considered "modern."[1]

"The father of modern chemistry" was not a scientist who dabbled in alchemy on the side. Boyle had a remarkably unified worldview, a worldview in which truths in chemistry, physics, alchemy, and theology were but different aspects of one unified body of knowledge. Boyle himself was quite aware that the individual aspects of his thought were

---

1. Most of the essays in *Robert Boyle Reconsidered,* edited by Michael Hunter (Cambridge: Cambridge University Press, 1994), reflect this trend; those relevant to this study will be cited individually in due course. See also Michael Hunter, "Alchemy, Magic and Moralism in the Thought of Robert Boyle," *British Journal for the History of Science* 23 (1990), pp. 387–410; idem, "Casuistry in Action: Robert Boyle's Confessional Interviews with Gilbert Burnet and Edward Stillingfleet, 1691," *Journal of Ecclesiastical History* 44 (1993), pp. 80–98; idem, "The Conscience of Robert Boyle: Functionalism, 'Dysfunctionalism' and the Task of Historical Understanding," in *Renaissance and Revolution: Humanists, Scholars, Craftsmen and Natural Philosophers in Early Modern Europe,* edited by J.V. Field and Frank A.J.L. James (Cambridge: Cambridge University Press, 1993), pp. 147–159; Lawrence M. Principe, *Aspiring Adept: Robert Boyle and his Alchemical Quest* (Princeton: Princeton University Press, forthcoming).

but a part of a unified whole, and expressed his conception of the whole in *The Excellency of Theology* (1674):

> The gospel comprises indeed, and unfolds the whole mystery of man's redemption, as far forth as it is necessary to be known for our salvation: and the corpuscularian or mechanical philosophy strives to deduce all the phænomena of nature form [sic] adiaphorous matter, and local motion. But neither the fundamental doctrine of Christianity, nor that of the powers and effects of matter and motion, seems to be more than an epicycle (if I may so call it) of the great and universal system of God's contrivances, and makes but a part of the more general theory of things, knowable by the light of nature, improved by the information of the scriptures: so that both these doctrines, though very general, in respect of the subordinate parts of theology and philosophy, seem to be but members of the universal hypothesis, whose objects I conceive to be the nature, counsels, and works of God, as far as they are discoverable by us (for I say not to us) in this life.[2]

In this book, I trace one aspect of Boyle's thought – his views on the limits of human understanding – and show the extent to which this aspect, as a part of the "universal hypothesis," is reflected in his theological writings and in his writings on natural philosophy. The significance of Boyle's views on the limits of human understanding should not be underestimated. As one of the forerunners of so-called "British empiricism,"[3] Boyle had a tremendous influence. He was slightly older than, although a contemporary of, the first of the "great thinkers" in the Locke-Berkeley-Hume trio of British empiricism. To a greater extent than was Locke, Berkeley, or Hume, Boyle was a practicing natural philosopher and one who published extensively in natural philosophy; in fact, in the seventeenth century, Boyle's status as a natural philosopher was exceeded only by that of Isaac Newton, on whom he had a considerable influence.[4]

2. Boyle, *Excellency of Theology, Works*, vol. 4, p. 19.
3. For the reason why British empiricism is "so-called," see David Fate Norton, "The Myth of British Empiricism," *History of European Ideas* 1 (1980), pp. 331–344.
4. Betty Jo Teeter Dobbs, "Problems in Newton's Early Chemistry," in *Religion, Science, and Worldview: Essays in Honor of Richard S. Westfall*, edited by Margaret J. Osler and Paul Lawrence Farber (Cambridge: Cambridge University Press, 1985), pp. 18–22; idem, *The Foundations of Newton's Alchemy, or "The Hunting of the Greene Lyon"* (Cambridge: Cambridge University Press, 1975), *passim*. It is surprising that there has not been more scholarly attention devoted to the relationship of Boyle's thought

In the chapters that follow, I show how Boyle justified his empiricism by appealing to the limits of human understanding. I also show that Boyle's views on the limits of human understanding were determined by his belief that God, in choosing to limit the intellectual capacities of the rational products of his creation, set quite definite limits on the ability of human beings to comprehend that creation.

My study is a contextual one. It is so first because I believe that a scholar often finds that an understanding of the context within which a particular thinker expresses his or her views enhances the scholar's understanding of those views. Indeed, in many cases, a knowledge of the context is an important prerequisite for interpreting the thinker's text correctly. Boyle's *Discourse of Things above Reason* is an excellent example of how knowledge of the context can facilitate a correct interpretation of the text. In that work, Boyle repeatedly stated that no scriptural revelation is "really" contrary to reason (that is, implies a contradiction or is contradictory to some other known truth). A scholar can either take such assertions at face value – as did a recent scholar when he claimed that Boyle's distinction between things that are above reason and things that are contrary to reason "provided the middle course for the natural philosopher and man of science: denying value to contradiction, but bolstering faith by admitting that some things were truly above men's comprehension."[5] Or a scholar can delve more deeply into both the text and the context, in which case the scholar discovers that what Boyle was actually asserting is that nothing that is revealed in scripture is contrary to *God's* infinite reason (although a revelation may indeed be contrary to finite human understanding). Although ambiguities in Boyle's text may make a correct interpretation difficult, a study of that text within the context of the writings of Boyle's contemporaries on the same subject leaves little or no doubt as to the correct interpretation.

My study is contextual for a second reason. It is often the case that

---

to that of Newton. David Fate Norton has examined that relationship in "Hume and the Experimental Method"; I thank Professor Norton for permitting me to read this unpublished essay. I compare Boyle's and Newton's views on the competence of human reason in "Pursuing Knowledge: Robert Boyle and Isaac Newton," in *The Canonical Imperative: Rethinking the Scientific Revolution*, edited by Margaret J. Osler, forthcoming.

5. John Redwood, *Reason, Ridicule and Religion: The Age of Enlightenment in England 1660–1750* (Cambridge: Harvard University Press, 1976), p. 211. I discuss similar misinterpretations of Boyle's views on "things above reason" in the Conclusion.

particular issues debated in the long-ago past seem, in the eyes of many twentieth-century scholars, to be outdated and to have no relevance to the questions being studied today. For example, the question of whether in theology there are revealed truths that are above reason (or even contrary to reason) is most likely of interest today to only a few of the philosophers who specialize in the philosophy of religion. Although there are many more scholars who are interested in the historical relationship(s) between science and religion, it is still the case that relatively few scholars have an immediate concern with the topics covered in the present study. However, a study of the context within which Boyle developed his views on the limits of human understanding has implications for scholars involved in almost any area of inquiry, because the extent to which a thinker's views in one field are influenced by that thinker's (often unexamined) presuppositions in another field is a question of universal concern.

An examination of the ways in which a particular thinker acquires certain beliefs and the ways in which those beliefs, once acquired, can influence that thinker's *other* beliefs is pertinent to studies in any field and to the philosophical questions asked in any century. In short, I think that contextual studies contribute to a recognition of the ways in which our own preconceptions of the way the world is (or the way we think it should be) influence our own beliefs.

Any scholar must set limits to his or her work in order to keep the project manageable. In this book, I have concentrated on Boyle's views on the limits of human reason, emphasizing the contemporary theological controversies that provided the context for the development of those views and the ways in which those views were given concrete application in his own studies in theology and in natural philosophy. To limit the study in this way is not to imply that other contexts and other influences did not contribute significantly to Boyle's worldview. I do, however, leave the elucidation of those other contexts and other influences to other scholars.[6]

6. For Bacon's influence, see Rose-Mary Sargent, "*The Diffident Naturalist: Robert Boyle and the Philosophy of Experiment*" (Chicago: University of Chicago Press, 1995); idem, "Learning from Experience: Boyle's Construction of an Experimental Philosophy," in *Robert Boyle Reconsidered*, pp. 57–78. Margaret J. Osler has examined Gassendi's influence on Boyle in "The Intellectual Sources of Robert Boyle's Philosophy of Nature: Gassendi's Voluntarism and Boyle's Physico-Theological Project," in *Philosophy, Religion, and Science, 1640–1700*, edited by Richard Kroll, Richard Ashcraft, and Perez Zagorin (Cambridge: Cambridge University Press, 1992),

Sources, Style and Abbreviations

For Boyle's works I have relied on *The Works of the Honourable Robert Boyle*, edited by Thomas Birch, 6 vols. (London, 1772), reprinted with an introduction by Douglas McKie (Hildesheim: Georg Olms, 1965). This edition is the one most often used by Boyle scholars because of its wide availability and because a study of the original editions of Boyle's writings does not seem to add to our understanding of his thought.[7]

In using Boyle's works to document the claims set forth in this book, I have juxtaposed quotations from quite different periods in Boyle's life. This raises the question whether it is appropriate to do so, or whether Boyle's views on reason's limits as depicted here reflect his thought at only a particular stage in his life, especially since all of the essays containing sustained arguments on reason's limits were published during the last decade of his life.[8] Nevertheless, the views he expressed in these works are consistent with views he expressed piecemeal in earlier works. While Boyle's views on reason's limits may have undergone some minor changes of emphasis during the course of his lifetime (a topic I discuss in chapter 4), there were no substantive changes.[9] So it is legitimate to use quotations from his earlier works to flesh out his views as expressed in his more sustained discussions.

---

pp. 178–198. For influences from the alchemical tradition, see William R. Newman, "Boyle's Debt to Corpuscular Alchemy," in *Robert Boyle Reconsidered*, pp. 107–118, as well as Principe, *Aspiring Adept*. For the influence of Renaissance humanism, see the Introduction by John Harwood, editor, in *The Early Essays and Ethics of Robert Boyle* (Carbondale and Edwardsville: Southern Illinois University Press, 1991), pp. xv–lxix.

7. See, for example, Peter Alexander, *Ideas, Qualities and Corpuscles: Locke and Boyle on the External World* (Cambridge: Cambridge University Press, 1985), p. 11; Timothy Shanahan, "God and Nature in the Thought of Robert Boyle," *Journal of the History of Philosophy* 26 (1988), p. 557 n. 31.

8. Specifically, *A Discourse of Things above Reason* and its accompanying *Advices in judging of Things said to Transcend Reason* (1681) and *Reflections upon a Theological Distinction*, which was published along with the first part of *The Christian Virtuoso* (1690).

9. Certainly there were no major changes after Boyle's early reaction to excessive rationalism and dogmatism, a reaction that contributed to his growing interest in natural philosophy. For this shift in Boyle's thought and its causes, see Michael Hunter, "How Boyle Became a Scientist," *History of Science* 33 (1995), pp. 59–103.

I cite Boyle's works by short title (providing a guide relating short titles to full titles in the bibliography), volume number, and page number(s). I have retained the spelling and capitalization of Birch's edition, and I have also reproduced exactly passages quoted from other authors; the only exceptions to this are that I have changed the initial letter of quotations to upper or lower case and I have altered the final punctuation of quoted passages when doing so has preserved the syntax of my own text. I have deliberately refrained from using "sic" except where doing so indicates an error in the original or is necessary to understanding the text being quoted; archaic constructions and inconsistencies in spelling and capitalization reflect the originals. For the convenience of modern readers, when quoting from manuscripts I have systematically given the modern forms of the minuscule 'u'/'v', 'i'/'j', and of the majuscule 'I'/'J'.[10] Similarly, I have silently expanded standard abbreviations, and I have expanded the thorn to 'th' throughout.

Concerning dates, I have assumed that the new year began on January 1 rather than on March 25. This was the common practice in the second half of the seventeenth century in England in contexts other than specifically legal ones. In the interests of consistency, I have modernized dates earlier in the century as well.[11]

In citing passages from scripture, I have used the King James Version (1611).

10. Harwood has noted that consistently modernizing Boyle's handwriting makes him appear more "modern" than he was (*The Early Essays and Ethics of Robert Boyle*, p. lxvii). However, by consistently rendering individual letters of the alphabet in their archaic form, even in those cases where Boyle himself had slipped into modern practice, Harwood makes Boyle appear *less* modern than he was.

11. For the differences between the Gregorian and Julian calendars, see Richard S. Westfall, *Never at Rest: A Biography of Isaac Newton* (Cambridge: Cambridge University Press, 1980; first paperback edition, 1983), p. xvi.

Acknowledgments

This book has been almost ten years in the making. It grew out of my graduate studies at the University of Kentucky, and I am very grateful for the valuable advice and encouragement of Dan Breazeale, Bruce Eastwood, Don Howard, Tom Olshewsky, John Shawcross, and, most especially, Jim Force and Henry Schankula. I cannot imagine having had a more helpful or supportive dissertation committee.

The last two drafts of this book were written here at Auburn University, and I thank Charlie Brown and Delos McKown for their support of the project. I am also deeply grateful for John Heilman's assistance in obtaining funding to support the revision process.

I want to express a very special "thank you" to Michael Hunter, who served as my on-the-spot advisor while I pursued my research in England, and who facilitated that research in so many ways that to enumerate them would require a separate volume. Since then, Michael has remained a constant source of ideas, inspiration, archival material, and news from the world of Boyle studies. Another heartfelt "thank you" is due Maggie Osler, who, via e-mail, offered almost daily advice, suggestions, and critical comments as I struggled to turn the dissertation into a book. A glutton for punishment, Maggie not only commented on one complete draft, but without complaint read (and reread) subsequent drafts of a number of chapters. More than once her enthusiasm for the project and her encouragement provided me with the motivation to continue plodding through the revision process. Few scholars are as generous with their time and support as Michael and Maggie have been. Having had the opportunity to get to know them and work with them has been a privilege.

This book has been greatly improved by suggestions made by two anonymous readers for Cambridge University Press, as well as by those of Ted Davis (who contributed significantly to chapter 8) and Larry Principe (who came to the rescue with chapter 5). In addition, I have profited from comments and suggestions made by participants at conferences and seminars where I read portions of various chapters.

A number of scholars have permitted me to read works in progress. For this I am grateful, and thank Mike Barfoot, Justin Champion, Ted Davis, Jim Force, Michael Hunter, Ron Levy, Jack MacIntosh, David Fate Norton, Maggie Osler, Larry Principe, and Rose-Mary Sargent.

Because I, like Boyle, suffer from an inability to bring a sentence to a successful conclusion, readers will be as grateful as I am for the superb job done by Ronald Cohen, my manuscript editor. Working with him has been a pleasure, and I have learned much in the process.

My research would have been impossible without the assistance of librarians at the British Library, the Bodleian, the Royal Society of London, Dr. Williams' Library, the library of Birkbeck College (University of London), the University of Kentucky, and Auburn University (where Harley Brooks was particularly helpful). In each case, the patience of those on whose expertise I depended was exceeded only by their competence.

Funding for research and writing has been provided by the Foundation for Intellectual History, the Graduate School at the University of Kentucky, the College of Liberal Arts at Auburn University, and the Research-Grants-in-Aid program at Auburn University. Funding for the research for my dissertation, without which the entire project would never have gotten off the ground, was provided by the American Association of University Women.

Others to whom I want to express my appreciation for various kindnesses include Gill Barfoot, Carolyn Brown, Astrid Force, Bob Genthner, and Dick and Julie Popkin. Working with Alex Holzman at Cambridge University Press has been a genuine pleasure. I am especially grateful to my parents, Harris and Imogene Walton, and to my children, Kelly and her husband Craig, and Paul and his wife Char, for their patience and understanding during the long periods of time when my attention was focused on my work and my family got the short end of the stick.

Introduction

Robert Boyle as Lay Theologian

Robert Boyle's status as a lay theologian was recognized in the seventeenth century and has been acknowledged ever since. In the sermon preached at Boyle's funeral in 1692, Gilbert Burnet, Bishop of Salisbury and one of Boyle's confessors, characterized him as one of those individuals who

> have directed all their enquiries into Nature to the Honour of its great Maker: And have joyned two things, that how much soever they may seem related, yet have been found so seldom together, that the World has been tempted to think them inconsistent; A constant looking into Nature, and a yet more constant study of Religion, and a Directing and improving of the one by the other.[1]

In 1701, Jeremy Collier, in his *Great Dictionary*, placed more emphasis on Boyle as a lay theologian than on Boyle as a natural philosopher.[2] And in his *Memoirs of the Lives and Characters of the Illustrious Family of the Boyles*, Eustace Budgell described the relationship between Boyle's theological writings and his scientific writings much as had Burnet, noting that he had "often blended *Religion* and *Philosophy* happily enough together; and made each serve to illustrate and embellish the other."[3] Similar characterizations of Boyle can be found in *Remarks on*

---

1. Gilbert Burnet, *A Sermon Preached at the Funeral of the Honourable Robert Boyle at St. Martins in the Fields, January 7, 1691/2* (London, 1692), p. 8. Burnet's sermon is now available in *Robert Boyle by Himself and His Friends, with a fragment of William Wotton's lost "Life of Boyle,"* edited by Michael Hunter (London: William Pickering, 1994); the passage quoted is on p. 39.
2. Jeremy Collier, *The Great Historical, Geographical, Genealogical and Poetical Dictionary*, 2nd edition, 2 vols. (London, 1701), s.v. "Robert Boyle" (alphabetized under "R").
3. Eustace Budgell, *Memoirs of the Lives and Characters of the Illustrious Family of the Boyles; Particularly of the Late Eminently Learned Charles Earl of Orrery. . . . With a Particular Account of the famous Controversy*

the Religious Sentiments of Learned and Eminent Laymen (London, 1792), in Henry Rogers's Introductory Essay to his edition of Boyle's *Treatises on The High Veneration Man's Intellect owes to God; On Things Above Reason; and on The Style of the Holy Scriptures* (1835), and in Richard B. Hone's *The Lives of Nicholas Ridley, D.D., Bishop of London; Joseph Hall, D.D., Bishop of Norwick; and The Honourable Robert Boyle* (1837), volume 3 in the Lives of Eminent Christians series.

Boyle himself was well-aware of his status as a lay theologian. The fact that approximately half of his voluminous writings deal with theological matters speaks for itself, as does the fact that in many of his works, theological concerns are so interwoven with his thoughts on natural philosophy that it is impossible to classify some works as either primarily theological or as concerned primarily with natural philosophy.[4] Further,

---

between the Honourable Mr. Boyle, and the Reverend Dr. Bentley, concerning the Genuineness of PHALARIS'S Epistles; also the same translated from the Original Greek. With an Appendix Containing the Character of the Honourable Robert Boyle Esq; Founder of an Annual Lecture in Defence of Christianity. By Bishop Burnet, and others. Likewise his Last Will and Testament (London, 1737), p. 126. Budgell's work was admittedly an apologetic one; his goal was to defend Charles Boyle, Robert Boyle's nephew, in Charles's controversy with Richard Bentley; for a discussion of the controversy see Joseph M. Levine, The Battle of the Books: History and Literature in the Augustan Age (Ithaca: Cornell University Press, 1991), esp. pp. 47–84. Nevertheless, his characterization of Robert Boyle as expressed in the phrase quoted is accurate.

4. Probably the best examples of works in which Boyle interwove theological considerations with his concerns in natural philosophy are "Essay IV" in *The Usefulness of Experimental Natural Philosophy* ("Containing a requisite Digression concerning those, that would exclude the Deity from intermeddling with Matter," published in 1663), *Some Considerations about the Reconcileableness of Reason and Religion,* to which Boyle appended *Some Physico-theological Considerations about the Possibility of the Resurrection* (1675), *A Discourse of Things above Reason* and its accompanying *Advices in judging of Things said to transcend Reason* (1681), *A Free Inquiry into the Vulgarly Received Notion of Nature* (1686), *A Disquisition on the Final Causes of Natural Things* (1688), and *The Christian Virtuoso* (1690), its *Appendix,* and its *Second Part* (1744). Boyle's earliest writings, only recently published, reveal an almost exclusive concern with matters of Christian morality and devotion; not until the early 1650s did Boyle's interest in natural philosophy become anything more than a generalized curiosity. See the Introduction by John Harwood, editor, in *The Early Essays and Ethics of Robert Boyle* (Carbondale and Edwardsville: Southern Illinois University Press, 1991), pp. xv-lxix; Michael Hunter, "How Boyle Became a Scientist," *History of Science* 33 (1995), pp. 59–103.

Boyle thought that *because* he was a layman his theological writings would be taken more seriously than they would be if he were a clergyman, and claimed that

> when the vessel [of religion] is threatened with shipwreck, or boarded by pirates, it may be the duty, not only of professed seamen, but any private passenger, to lend his helping hand in that common danger. And I wish I were as sure, that my endeavours will prove successful, as I am, that such churchmen, as I most esteem, will think them neither needless nor unseasonable. Nay, perhaps my being a secular person may the better qualify me to work on those I am to deal with, and may make my arguments, though not more solid in themselves, yet more prevalent with men, that usually (though how justly, let them consider) have a particular pique at the clergy, and look with prejudice upon whatever is taught by men, whose interest is advantaged by having what they teach believed.[5]

Twentieth-century scholars have continued the tradition of seeing Boyle as a lay theologian. Scholars have, for example, investigated the relationship between Boyle's voluntarism and his empirical scientific methodology, as well as the similarities in his approaches to God's two books (the book of nature and the book of scripture). Attention has been paid to his ethical writings, his views on spirit-contact, and the ways in which his piety affected his personality. His alchemical pursuits have been scrutinized in the light of his theological concerns, as have his views on the limits of mechanism in the corpuscular philosophy.[6]

5. Boyle, *Reason and Religion, Works,* vol. 4, p. 153.
6. For Boyle's voluntarism, see the sources cited in chapter 8 of this book, n. 45. For his approach to the "two books" see Rose-Mary Sargent, *The Diffident Naturalist: Robert Boyle and the Philosophy of Experiment* (Chicago: University of Chicago Press, 1995), esp. pp. 109–128. For his ethical writings, see *Early Essays,* edited by Harwood. For spirit-contact and piety, see Michael Hunter, "Alchemy, Magic, and Moralism in the Thought of Robert Boyle," *British Journal for the History of Science* 23 (1990), pp. 387–410; idem, "Casuistry in Action: Robert Boyle's Confessional Interviews with Gilbert Burnet and Edward Stillingfleet, 1691," *Journal of Ecclesiastical History* 44 (1993):80–98. For alchemy and theological concerns, see (in addition to the two essays by Hunter cited immediately above), Lawrence M. Principe, "Boyle's Alchemical Pursuits," in *Robert Boyle Reconsidered,* edited by Michael Hunter (Cambridge: Cambridge University Press, 1994), pp. 91–105; see also the first comprehensive treatment of Boyle's preoccupation with transmutational alchemy in idem, *Aspiring Adept: Robert Boyle and his Alchemical Quest* (Princeton: Princeton University Press, forthcoming). For the limits of mechanism, see John Henry, "Boyle and Cosmical Qualities," in *Robert Boyle Reconsidered,* pp. 119–138.

Despite the wide variety of topics related to Boyle's theological views discussed by scholars, most of the emphasis in the secondary literature, at least until very recently, has been on his natural theology.[7] Far from considering the new science as posing a threat to religion, Boyle thought that the natural philosopher was in a far better position to appreciate the arguments of natural religion than were most other people. The "virtuoso" (or one who "understands and cultivates experimental philosophy") was, Boyle thought, in a unique position to gather "experience ... on which he is disposed to make such reflections, as may (unforcedly) be applied to confirm and encrease in him the sentiments of natural religion, and facilitate his submission and adherence to the Christian religion." Although "in almost all ages and countries ... perfunctory considerers" are led by a consideration of design in the universe to assent to the basic truths of natural religion (which are that God exists, that we can infer some of his attributes, and that the human soul is immortal), the assent of such "perfunctory considerers" is inferior to the assent given by the natural philosopher.[8]

There are two particular ways in which, in Boyle's opinion, the study of natural philosophy facilitates the acceptance of the truths of natural religion. First, the natural philosopher studies final causes, and from the consideration of the power and wisdom of the creator as displayed in the creation is led to acknowledge God's existence. Second, the natural philosopher learns to distinguish material from immaterial substances and comes to realize that body and soul cannot have the same essential attributes. The natural philosopher, observing that bodies are perishable, can infer that souls are not.[9]

God's providence (especially as expressed in final causes) and the imperishability of the soul serve as Boyle's "bridge" from natural to

7. The best account remains Richard S. Westfall, *Science and Religion in Seventeenth-Century England* (New Haven: Yale University Press, 1958). See also R.M. Burns, *The Great Debate on Miracles: From Joseph Glanvill to David Hume* (Lewisburg, PA: Bucknell University Press, 1980); Harold Fisch, "The Scientist as Priest: A Note on Robert Boyle's Natural Philosophy," *Isis* 44 (1953), pp. 252–265; M.S. Fisher, *Robert Boyle, Devout Naturalist: A Study in Science and Religion in the Seventeenth Century* (Philadelphia: Oshiver Studio Press, 1945); and L.T. More, *The Life and Works of the Honourable Robert Boyle* (New York: Oxford University Press, 1944).
8. Boyle, *Christian Virtuoso, Works*, vol. 5, pp. 513–524; the quotations are from pp. 513, 524, and 516 respectively.
9. Boyle, *Christian Virtuoso, Works*, vol. 5, pp. 515–522.

revealed religion. God would not have left human beings without the means to obtain the true end of their imperishable souls – eternal happiness in heaven – so each individual must assume that God has in some way revealed what must be believed and done in order to reach that end. And, of course, Boyle believed that there has been just such a revelation concerning the worship of and obedience to God necessary for salvation – the Christian religion.[10]

Boyle was aware, of course, that Christianity was not the only religion claimed by its adherents to have been revealed by God. Therefore, before assenting to the propositions of the Christian revelation, individuals must judge that it is indeed the only genuine such revelation. Here again, reason enters the picture. Boyle offered two reasons for accepting the Christian revelation as divinely inspired. First was the excellency of the doctrine. Second, God had attested to its truth by performing miracles (including the miracle of Christianity's rapid spread, which had been prophesied).[11] Although miracles were a violation of the uniformity of nature as established by God, it was not irrational, Boyle thought, to believe in them when God's omnipotence was taken into consideration; further, that miracles had occurred was testified to by individuals of unimpeachable motives. Human reason is capable of judging these signs and evaluating each as evidence for the authority of the Christian revelation.[12]

10. Boyle, *Christian Virtuoso, Works,* vol. 5, p. 522; the term "bridge" is Boyle's.

11. Boyle considered prophecies to be a species of miracle, claiming that "true prophecies of unlikely events, fulfilled by unlikely means, are supernatural things; and as such … may properly enough be reckoned among miracles" (*Christian Virtuoso, Works,* vol. 5, p. 535). This connection between miracles and fulfilled prophecies would be noted later by David Hume in his essay "On Miracles," in *An Enquiry Concerning Human Understanding* (*Enquiries Concerning Human Understanding and Concerning the Principles of Morals,* reprinted from the posthumous edition of 1777, edited by L.A. Selby-Bigge, 3rd edition, with text revisedand notes by P.H. Nidditch [Oxford: Clarendon Press, 1975], p. 130).

12. See, for example, Boyle, *Christian Virtuoso, Works,* vol. 5, pp. 522, 524, and 531; *Appendix* to *Christian Virtuoso, Works,* vol. 6, p. 677; and *Reason and Religion, Works,* vol. 4, p. 162. Boyle also argued for the unique truth of the Christian religion in his unpublished essay, "De Diversitate Religionum" (Boyle Papers, vol. 6, fols. 279–291). This work exists in Latin translation only; it will be included along with an English translation in the forthcoming Pickering edition of Boyle's *Works,* edited by Michael

This portrayal of Boyle as one who stressed the *reasonableness* of Christianity by arguing that human reason, unaided by revelation, is capable of discerning the truths of natural religion upon which a belief in the Christian revelation can be based is correct, but only partially so, and the emphasis on this aspect of Boyle's thought in the secondary literature is unfortunate. First, it has led to the characterization of Boyle as being one of those virtuosi from whose writings "deism, the religion of reason, steps full grown."[13] Such a characterization is incorrect for two reasons. It has obscured the extent to which Boyle's writings on "things above reason" distinguish him from such virtuosi as Joseph Glanvill and John Locke, whose writings on the reasonableness of Christianity were not qualified, as were Boyle's, by any extended discussion of revelations that were impervious to reason.[14] Second, scholars have only recently realized the extent to which deism sprang from the writings of individuals (such as the Socinians) who, unlike Boyle, insisted that scrip-

Hunter and Edward B. Davis. The inclusion of this and other such manuscripts, a project supervised by Hunter, is being funded by the Leverhulme Trust.

13. Westfall, *Science and Religion*, p. 219. Westfall's account as a whole is more balanced than this quotation indicates. For example, he notes correctly that Boyle sets quite definite limits on the competence of human reason to judge the "superior truths of Christianity" (p. 174). Nevertheless, Westfall failed to pursue this aspect of Boyle's thought. Similarly, Leroy E. Loemker portrayed Boyle as contributing to the spread of deism ("Boyle and Leibniz," *Journal of the History of Ideas* 16 [1955], pp. 22–43). For other studies on the relationship of the seventeenth-century emphasis on natural theology and the rise of English deism, see Robert E. Sullivan, *John Toland and the Deist Controversy: A Study in Adaptations* (Cambridge: Cambridge University Press, 1982); Stephen H. Daniel, *John Toland: His Methods, Manners, and Mind* (Kingston and Montreal: McGill-Queen's University Press, 1984); John Orr, *English Deism: Its Roots and Fruits* (Grand Rapids, MI: Eerdmans, 1934); and Leslie Stephen, *History of English Thought in the Eighteenth Century*, 3rd edition, vol. 1 (New York: G.P. Putnam's Sons, 1902).

14. I discuss Glanvill's views on reason and revelation in chapter 2. For Locke's views, see John Marshall, "John Locke and Latitudinarianism," in *Philosophy, Science, and Religion in England 1640–1700*, edited by Richard Kroll, Richard Ashcraft, and Perez Zagorin (Cambridge: Cambridge University Press, 1992), pp. 253–282; idem, *John Locke: Resistance, Religion and Responsibility* (Cambridge: Cambridge University Press, 1994); Gerard Reedy, S.J., *The Bible and Reason: Anglicans and Scripture in Late Seventeenth-Century England* (Philadelphia: University of Pennsylvania Press, 1985), pp. 119–141.

ture must be interpreted in such a way that the *content* of revelation be consonant with human reason.[15]

Of greater significance is the fact that this emphasis on the role of natural theology in Boyle's thought has resulted in the neglect of Boyle's emphatic denial that human reason is competent to judge the content of revelation. This neglect is particularly regrettable because of the close affinity in Boyle's thought between his views on the limits of human understanding in the context of revealed religion and his views on the limits of human understanding in the context of the natural philosopher's quest to understand the secrets of nature. In this book, I explore that affinity and argue that Boyle's views on reason's limits affected his conception of the proper goals and methodology of the new natural philosophy. Further, I argue that Boyle's theological beliefs provided the foundation for his views on natural philosophy: Boyle believed that God, in creating human beings, deliberately limited reason's power and scope. It was this (essentially unexamined) starting point from which his arguments concerning the limits of reason's competence followed.

In Part I, I examine the theological context within which Boyle developed his views on things above reason. In chapter 1, I survey briefly the history of various conceptions of the proper relationship of reason to religion from the beginning of Christianity to Boyle's era, with an emphasis on the concepts that are particularly relevant for an understanding of his thought. In chapter 2, I investigate the claim of the Socinians that scriptural revelation should be interpreted in such a way as to be consonant with human reason, as well as responses to that claim made by some of Boyle's contemporaries. Socinian ideas were spreading rapidly in England in the 1650s, and many of the arguments that Boyle made in his *Discourse of Things above Reason* (1681) can be traced back to the arguments of various nonconformists concerning reason's limits in the ensuing controversies. These themes, in turn, can be traced back to their medieval origins. Instead of aligning himself with latitudinarian contemporaries, such as Joseph Glanvill, who emphasized reason's competence, Boyle aligned himself with those who more stringently circumscribed reason's role in understanding revelation.

If the nonconformists tended to stress reason's limits in their refutations of Socinian doctrines and methodology, certain of them (and most

15. For the relationship between Socinianism and deism, see especially J.A.I. Champion, *The Pillars of Priestcraft Shaken: The Church of England and its Enemies 1660–1730* (Cambridge: Cambridge University Press, 1991); Sullivan, *John Toland and the Deist Controversy*; Reedy, *The Bible and Reason*, esp. pp. 119–141.

especially the high Calvinists among them) dogmatically asserted that *their* interpretations of incomprehensible doctrines were uniquely correct. In short, they claimed to have comprehended correctly the incomprehensible, and contributed volume after volume to the heated doctrinal debates in seventeenth-century England. One of the most intense of those controversies was the question of the proper interpretation of the doctrine of predestination. In chapter 3, I survey the issues involved in the predestinarian controversies, with an emphasis on those aspects of particular concern to Boyle. I end chapter 3 with an examination of John Howe's *Reconcileableness of God's Prescience of the Sins of Men, with the Wisdom and Sincerity of his Counsels, Exhortations, and whatsoever Means He uses to prevent them* (1677), which he had written, he said, "at the request of Mr. Boyle," as well as the controversy which the work generated. Boyle's *Things above Reason* should be read and interpreted in the context of both the religious rationalism urged by the Socinians and the controversies over predestination between the Calvinists and Arminians.

In chapter 4, I examine *Things above Reason* and its accompanying *Advices in judging of Things said to transcend Reason*, incorporating material from Boyle's related writings when relevant. In this chapter, I describe his categories of things that are above or even contrary to reason because they are either incomprehensible, inexplicable, or unsociable, and argue that in Boyle's view, human reason is so incompetent to judge the content of revelation that even the law of noncontradiction may appear to be violated from the perspective of finite human understanding. In addition, I show that Boyle was aware that his emphasis on reason's limits invited charges of enthusiasm, that he anticipated this objection, and answered it in *Advices*. Further, I argue that *Things above Reason* and *Advices* were written in response to the fervent polemics characteristic of theological controversies in his day, Boyle's argument being that if a doctrine is truly above the ability of human understanding to comprehend (which the controversialists themselves acknowledged), then any pretensions to have attained a uniquely correct understanding of that doctrine must be abandoned.

In Part II, I turn to the question of the relationship between Boyle's views on the limits of reason in theology to his conception of the task of the natural philosopher. After discussing briefly each of the major theories of matter he considered to be viable alternatives to his preferred corpuscular hypothesis (in chapter 5), I turn to Boyle's views on the various possible sources of knowledge of the created world in chapter 6. Specifically, I discuss his views on scriptural revelation, personal revelation, abstract reason (including innate ideas, which, having their origin in God, might be construed as a form of revelation), and sensory perception as sources of knowledge of the created world.

I begin chapter 7 with an examination of Boyle's three categories of things above reason in the context of natural philosophy, showing that he believed that just as there are incomprehensible, inexplicable, and unsociable truths in theology, he also believed that there are incomprehensible, inexplicable, and (from a practical although not from a theoretical point of view) unsociable truths in the field of natural philosophy. I then turn to Boyle's evaluations of each of the alternative theories of matter I described in chapter 5. I use his criteria for good and excellent hypotheses to argue that his conception of the proper goal of the natural philosopher was that the naturalist should provide *intelligible* explanations (not, necessarily, true explanations) of phenomena that themselves were often incomprehensible and inexplicable. Further, I argue that his conception of the limits of human reason kept him from declaring straightforwardly that the viable alternative theories were false, even though he made it clear that he rejected them. In short, I stress the provisional nature of Boyle's claims. In addition, I emphasize that he believed that some of nature's secrets could not be explained intelligibly at all, and that although he thought progress could be expected in the investigation of nature's secrets, he did not think that human understanding would ever be able to penetrate all of them (at least not in this life).

Although in chapter 7 I discuss the theological concerns that lay beneath many of Boyle's objections to the viable alternative theories of matter, my emphasis is on objections he considered to be totally secular in nature. My point in doing so is to show that even when Boyle *thought* he had excluded any specific theological considerations from a given argument, he was in fact *always* assuming that human reason is extremely limited in its power and scope, and that this assumption itself was based on his voluntaristic conception of God. In chapter 8, I examine Boyle's voluntarism, and argue that he believed that God, in creating human beings, had freely chosen to limit the power and scope of human reason. In creating the world commensurate to his own infinite understanding and in limiting the rational faculties of created beings, God had deliberately left "human understandings to speculate as well as they could upon those corporeal, as well as other things."[16] Boyle believed that by limiting human understanding in this life, God had wisely re-

16. Boyle, *Appendix* to *Christian Virtuoso*, *Works*, vol. 6, p. 694.

served a full and complete understanding of both the secrets of theology and of nature for the next life, thereby providing the greatest possible reward for being both a Christian and a virtuoso while here on earth.

Before proceeding, it will be helpful to smooth the way for the chapters in Part I by discussing Boyle's views on religious controversies, for my claim that his *Things above Reason* emerged out of the religious controversies of his day runs counter to the image we have of Boyle as having removed himself from those very controversies. In the remaining three sections of this Introduction, I discuss Boyle's early responses to religious controversies, his views on liberty of conscience, and the tension between his "love of peace" and his "love of truth."

Boyle's Early Responses to Religious Controversies

After his death, Boyle would be remembered as a natural philosopher and lay theologian who refrained from participating in the theological controversies of his day. Gilbert Burnet, Bishop of Salisbury, for example, emphasized Boyle's irenicism during the sermon he preached at his funeral. Burnet noted that Boyle disliked "any Nicety that occasioned Divisions amongst Christians," and that he "was much troubled at the Disputes and Divisions which had arisen about some lesser Matters." Religion, Boyle thought, ought to purify hearts and govern lives, and, according to Burnet, he

> avoided to enter into the unhappy Breaches that have so long weak-
> ened, as well as distracted Christianity, any otherwise than to have
> a great aversion to all those Opinions and Practices, that seemed to
> him to destroy Morality and Charity.[17]

In the first published biography of Boyle – the *Life* prefixed to the 1744 and 1772 editions of Boyle's *Works* – Thomas Birch quoted these

17. Burnet, *Sermon Preached at the Funeral*, pp. 25–27.Burnet's comment that Boyle had an aversion to any opinion or practice that destroyed charity most likely refers to Boyle's aversion to sectarian controversies themselves; Boyle's *Discourse of Things above Reason* (1681) was (at least partially) an attempt to persuade his fellow Christians to abandon participation in such controversies; see chapter 4. Burnet's comment about opinions and practices that destroy morality might refer generally to Christians living immoral lives or it might refer to a dislike on Boyle's part for antinomianism, a belief that strict Calvinists were often accused of holding (on the grounds that absolute predestination might lead to the belief that the elect cannot fall from grace regardless of what they do).

remarks from Burnet's funeral sermon.[18] Birch's comments, in turn, influenced subsequent biographers. In an introductory essay to an 1835 edition of three of his theological works, Henry Rogers, for example, stated that Boyle

> took no part in the unhappy controversies which distracted the age. His serene and placid spirit recoiled from controversies of every kind, but especially from such as were alike distasteful to his temper and alien from his pursuits, and which appeared to him, as they must to every other sober mind, to have been prosecuted with an animosity and rancour so utterly disproportionate to their importance.[19]

In one sense, this is a correct interpretation. In an era noted for its heated religious polemics, Boyle's writings reflect a conscious decision not to involve himself in sectarian debates. He did, in fact, have an aversion to the heated polemics of his day. However, a careful examination of his *Discourse of Things above Reason* (1681) in the context of the theological controversies of his day reveals that in that work he was indeed participating in not only one but two of the ongoing theological debates of seventeenth-century England – the question of the proper use of human reason in attempting to unravel the mysteries of Christianity, and the question of the proper interpretation of the doctrine of predestination. In fact, as I show in chapter 4, *Things above Reason,* with its dialogue format and its nondogmatic tone, was intended as a model of the *proper* way to debate the correct interpretation of scripture.

An examination of his early correspondence and theological writings reveals that in the years following his return to England from his studies abroad in 1644, Boyle struggled to formulate an appropriate response to the sectarianism then at its height. In doing so, he had to deal with a number of related issues. One of these was the extent to which liberty of conscience should be tolerated (if at all) in religious matters. Another was to what extent, if any, he should involve himself in doctrinal controversies.

Boyle was only twelve years old in 1639 when he left the British Isles to travel and study abroad. In 1644, he returned to an England deep in the throes of civil war, not only torn by political conflict but also riven by religious controversies. On arrival, he headed for London,

18. Thomas Birch, *The Life of the Honourable Robert Boyle, Works,* vol 1, p. cxli.

19. Henry Rogers, Introduction to Boyle's *Treatises on the High Veneration Man's Intellect owes to God, On Things Above Reason, and on the Style of the Holy Scriptures,* edited by Henry Rogers (London, 1835), p. xliii.

discovering when he arrived that his sister Katherine, Lady Ranelagh, was living there, having moved to London to escape the dangers of the Irish Rebellion.

He remained with his sister for more than four months, and while there he was no doubt brought up to date on the whereabouts and activities of other members of his family. Boyle's father, who had died the previous year, had remained loyal to the king, although there is some evidence in his private papers that he had had Parliamentarian and Puritan sympathies. His brother Lewis, Viscount Kinalmeaky, had died in battle in September of 1642 during the Irish Rebellion. His eldest brother Richard, now Earl of Cork, had married into a Royalist family and, as a representative of his wife's family, served in the Royalist army.

Katherine herself was a patron of Puritan scientists and divines and had close ties with the Hartlib Circle, the members of which were preparing for the millennium by reforming learning, unifying Christians, and converting the Jews. Another sister, Mary Rich, Countess of Warwick, would become a fully fledged Puritan whose constant preoccupation with virtue and piety is revealed in her unpublished diary.[20] Other family members, concerned with the defense of and recovery of their Irish properties, aligned themselves with whichever party was in power

20. British Library Add. MSS. 27,351-27,355. A typical entry begins: "In the morning as soon as I awaked I blessed God then went out alone into the wilderness to meditate, and there God was pleased to give me sweet communion with him, and to fix my thoughts much upon my death and to make me pray to God with strong cryes and abundance of tears, that I might be prepared for that great change, and then God was pleased to make me meditate upon the joys of heaven and to make me consider heaven would make me eternally happy in the fruition of God in love which did mightily [illeg.] my heart with desires to enter into this joy, my soul was exceedingly carried out in love to Christ's person, and with desires to be with him, and I came away much refreshed, and my heart exceedingly cheered, after I was drest I went into my closset, read, and then prai'd and there too the desires of my heart went out exceedingly after God, I blest God heartily for his mercies, went then to family prayer the heart breathed after God. . . ." [30 July 1666, MS. 27,351 fols. 11r-11v]. For Mary Rich's piety, see Sara Heller Mendelson, *The Mental World of Stuart Women: Three Studies* (Brighton, Sussex: Harvester Press, 1987), pp. 62-115. Michael Hunter has suggested that Boyle shared the "deep, agonised, piety" of his sister ("The Conscience of Robert Boyle: Functionalism, 'Dysfunctionalism' and the Task of Historical Understanding," in *Renaissance and Revolution: Humanists, Scholars, Craftsmen and Natural Philosophers in Early Modern Europe*, edited by J.V. Field and F.A.J. James [Cambridge: Cambridge University Press, 1993], pp. 147-59).

in England at any given time, and successively shifted allegiances from Charles I to parliament, from parliament to Cromwell, and from Cromwell to Charles II. The most prominent of these was Roger, Baron Broghill.[21]

Having relatives on both sides of the war left Boyle, as he noted in a letter of October 1646 to his former tutor Isaac Marcombes, "obnoxious to the injuries of both parties, and the protection of neither." As a result, he observed "a very great caution."[22] Further, he could not avoid being acutely aware of the political uncertainty and the hazards of civil war in his frequent travels between Stalbridge, the Dorset County manor where he lived from 1645 until he moved to Oxford in the mid-1650s, and London.[23] Obviously, he experienced the political sectarianism firsthand.

As I have already noted, when Boyle returned from his studies abroad,

21. For Boyle's family background, see Nicholas Canny, *The Upstart Earl: A Study of the Social and Mental World of Richard Boyle, first Earl of Cork, 1566–1643* (Cambridge: Cambridge University Press, 1982). The most complete account of Boyle's early life is found in R.E.W. Maddison, *The Life of the Honourable Robert Boyle F.R.S.* (London: Taylor and Francis, 1969). For Lady Ranelagh's Interregnum activities, see Charles Webster, *The Great Instauration: Science, Medicine and Reform, 1626–1660* (New York: Holmes and Meier, 1976), pp. 62–63. James R. Jacob's *Robert Boyle and the English Revolution: A Study in Social and Intellectual Change* (New York: Franklin, 1977) is less than reliable because it is based on selective readings of Boyle's texts and because Jacob perhaps erroneously assumes that attitudes expressed by some of Boyle's colleagues were also Boyle's. The best overall account of Boyle's interests and activities during this period is Hunter, "How Boyle Became a Scientist." For Boyle's political concerns during the Interregnum, see Malcolm Oster, "Virtue, Providence and Political Neutralism: Boyle and Interregnum Politics," in *Robert Boyle Reconsidered*, pp. 19–36. On the Hartlib Circle, see Richard H. Popkin, "The Third Force in 17th Century Philosophy: Scepticism, Science and Biblical Prophecy," *Nouvelles de la république des lettres* 3 (1983), pp. 35–63; Webster, *The Great Instauration*, pp. 32–50 and 67–77; and Maddison, *Life*, pp. 57–88, as well as the excellent essays in *Samuel Hartlib and Universal Reformation: Studies in Intellectual Communication*, edited by Mark Greengrass, Michael Leslie, and Timothy Raylor (Cambridge: Cambridge University Press, 1994).
22. Boyle to Marcombes, 22 October 1646, *Works*, vol. 1, p. xxxiii. Marcombes (d. c. 1654), a Frenchman by birth who made his home in Geneva, was the nephew by marriage of Jean Diodati, the strict predestinationist, and as such had connections with important Protestant families in England and on the Continent.
23. See Oster, "Virtue, Providence and Political Neutralism."

he found not only an England divided politically, but also an England which was splintered into religious factions.[24] In addition to the disestablished Anglicans, the Presbyterians, and the Independents, there were innumerable more radical sects, including Anabaptists, Familiasts, and Socinians; indeed, Boyle himself described the situation as being one in which there were "no less than 200 several opinions in point of religion."[25]

In the decade and a half after he returned to England, he struggled to formulate an appropriate response to this divisiveness; it would not be inaccurate to say that in this matter, too, Boyle came to exercise "a very great caution." In later chapters, I shall discuss his initial reactions to two particular controversies – the claims of the Socinians that human reason is the criterion against which differing scriptural interpretations should be judged, and the disputes between Arminians and Calvinists over the correct interpretation of the doctrine of predestination, for it was these two controversies that would ultimately prompt him to publish *Things above Reason*. First, however, I must note that no matter how Boyle might respond to particular doctrinal claims, that response would be conditioned by his respect for liberty of conscience. And his respect for liberty of conscience in turn was based, at least partially, on his belief that human beings cannot comprehend fully the content of God's revelation, a belief he would later use to argue that if God's revelation cannot be comprehended fully, then surely any claims of individuals to have arrived at a uniquely correct interpretation of that revelation must be suspect.

Liberty of Conscience

In the same October 1646 letter to Marcombes mentioned earlier, Boyle expressed his opinion that different religious views should be tolerated. Boyle had returned from his studies abroad, he complained, to find

24. For radical religious ideas during the Interregnum, see Christopher Hill, *The World Turned Upside Down: Radical Ideas during the English Revolution* (New York: Viking Press, 1972); B.S. Capp, *The Fifth Monarchy Men: A Study in Seventeenth-Century Millenarianism* (London: Faber, 1972); and the essays in J.F. McGregor and B. Reay, editors, *Radical Religion in the English Revolution* (New York: Oxford University Press, 1984).

25. I discuss the letter in which Boyle expressed this thought in the following section.

no less than 200 several opinions in point of religion, some digged out of those graves, where the condemning decrees of primitive councils had long since buried them; others newly fashioned in the forge of their own brains; but the most being new editions of old errors.[26]

Noting that an ordinance providing for the punishment of heretics was being considered, Boyle was of the opinion that toleration was a better response. After all, each of the "several opinions" was believed to be a revealed truth, and perhaps some of them were, for persons who "justly pretend to a greater moderation, suspect" that

our dotage upon our own opinions makes us mistake many for impostures, that are but glimpses and manifestations of obscure or formerly concealed truths, or at least our own pride or self-love makes us aggravate very venial errors into dangerous and damnable heresies.[27]

Even if some of the opinions were in fact heretical, an individual should not, Boyle thought, be punished for holding an erroneous religious opinion. Noting that the same individuals who urged hanging for heresy also argued that true beliefs are a work of divine revelation, Boyle stated that he could not understand why anyone would argue that an individual should be hanged because "it has not yet pleased God to give him his spirit." "Certainly to think by a halter to let new light into the understanding, or by the tortures of the body to heal the errors of the mind," Boyle continued, "seems to me like the applying a plaster to the heel, to cure a wound in the head."[28]

Boyle expressed much the same view about such erroneous opinions in a letter that he wrote in May 1647 to his friend and colleague, John Dury:

26. Boyle to Marcombes, 22 Oct. 1646, *Works*, vol. 1, p. xxxii.
27. Boyle to Marcombes, 22 Oct. 1646, *Works*, vol. 1, pp. xxxii-xxxiii. The ordinance Boyle had in mind was most likely a petition presented to Parliament by the Common Council of London in December 1646; for a discussion of its origin, contents, and fate, see Valerie Pearl, "London's Counter-Revolution," in *The Interregnum: The Quest for Settlement 1646–1660*, G.E. Aylmer, editor (London: Macmillan, 1972), pp. 41–44. Boyle was in London in October, when he wrote to Marcombes.
28. Boyle to Marcombes, 22 Oct. 1646, *Works*, vol. 1, p. xxxiii. Boyle believed that persuasion was the proper way to lead others to truth. For his rhetorical strategies, see John Harwood's Introduction to *Early Essays*, pp. liii-lxvii; Edward B. Davis, " 'Parcere nominibus': Boyle, Hooke and the Rhetorical Interpretation of Descartes," in *Robert Boyle Reconsidered*, pp. 157–175.

> As for our upstart sectaries (mushrooms of the last night's springing
> up) the worst part of them, . . . will be as sudden in their decay, as
> they were hasty in their growth; and indeed perhaps the safest way
> to destroy them is rather to let them die, than attempt to kill them.[29]

A number of factors might have influenced Boyle's views on liberty of
conscience. Perhaps he had been influenced at Eton, which he had
attended prior to his travels abroad, by the provost, Henry Wotton,
who had had inscribed on his tomb, "Disputandi Pruritus, Ecclesiarum
Scabies."[30] Perhaps during these years immediately following his return
from the Continent, Boyle was more directly influenced by John Dury.
The letter to Dury from which the earlier quotation was taken was
written to encourage Dury in his attempts to reconcile the Lutherans
and the Calvinists, and only a year before Boyle's letter to him, Dury
had published an English translation of the Socinian Johannes Crellius's
Vindiciæ pro religionis libertate. In this work, originally written in 1637,
Crellius had argued that Roman Catholics should grant religious liberty
to Protestants. Dury, in his preface, extended the argument to the situation in England and argued that Protestants should grant religious liberty
to other Protestants.[31]

29. Boyle to Dury, 3 May 1647, *Works*, vol. 1, p. xl.
30. Izaak Walton, *Life of Sir Henry Wotton* (London, 1670), in idem, *Lives of John Donne, Henry Wotton, Richard Hooker, and George Herbert* (London: George Routledge and Sons, 1888), p. 126.
31. Crellius's work was originally published under the pseudonym of Junius Brutus in "Eleutherapoli" (Amsterdam). Dury published it under the title of *A learned and exceedingly well-compiled Vindication of Liberty of Religion: Written by Junius Brutus in Latine, and Translated into English by N. Y. who desires, as much as in him is, to do good unto all men* (No Place [sic], 1646). I discuss the work in chapter 2. Boyle's sister Lady Ranelagh was actively supporting Dury and Hartlib before Boyle returned from Geneva in 1644 (Harwood, Introduction to *Early Essays*, p. xliv); it is likely that the two met through Lady Ranelagh shortly after Boyle's return to England in mid-1644. R.E.W. Maddison states that the *latest* date by which Dury was known to Boyle was 1645, which would be prior to the publication of Dury's translation of Crellius's work ("Studies in the Life of Robert Boyle, F. R. S.: Part VI, The Stalbridge Period, 1645–1655, and the Invisible College," *Notes and Records of the Royal Society of London*, 18 [1963], p. 105). For Boyle's having urged Dury to write an attempted reconciliation of Lutherans and Calvinists, see "Sir Peter Pett's Notes on Boyle," in *Robert Boyle by Himself and His Friends*, pp. 72–73; Malcolm Oster, "Millenarianism and the New Science: The Case of Robert Boyle," in *Samuel Hartlib and Universal Reformation*, pp. 146–147.

Perhaps Boyle's views on liberty of conscience were reinforced by the relative toleration of the Interregnum years. Although none of the various political factions was willing to extend toleration to the extent envisioned by Oliver Cromwell, the freedom to worship as one chose was the rule rather than the exception during these years, exceptions to this rule being the former Anglicans (the *Book of Common Prayer* having been proscribed) and Roman Catholics. And in the context of the 1660 Restoration, the question of toleration for Independents, Presbyterians, and members of the various radical sects was a crucial question. Indeed, two of Boyle's Anglican friends, Sir Peter Pett and Thomas Barlow, penned works at Boyle's request in which they argued that the reestablished Anglican church should extend toleration to most Protestants, although it should be noted that Boyle's concern for toleration seems to have been based on spiritual considerations rather than the more prudential considerations of Pett.[32]

In the October 1646 letter to Marcombes, Boyle had acknowledged the difficulty of knowing which, if any, alleged revelations of the Holy Ghost might be genuine. Quite possibly, Boyle's belief in liberty of

---

32. Thomas Barlow's "The Case of a Toleration in Matters of Religion," pp. 1–93 in *Several Miscellaneous and Weighty Cases of Conscience* (London, 1692), was written in 1660, when the question of religious toleration in the context of the Restoration was an issue. In it, Barlow argued that toleration should be granted if to deny it would threaten public peace (pp. 12–16). Toleration should be denied only to those who refused to acknowledge the supremacy of the civil magistrate – for example, some but not all papists, Quakers, and those whose beliefs violate nature, such as Adamites (pp. 29–35). Concerning others (and Barlow explicitly included Socinians), Barlow refused to declare that it is unlawful for the magistrate to use temporal punishment, but argued that the magistrate should not do so on the grounds (among others) that there is no warrant for such an approach in the gospel and that individuals should be persuaded of truth, not forced into it (pp. 36–62). For a discussion of Barlow's work, as well as Pett's *A Discourse Concerning Liberty of Conscience* (London, 1661), see James R. Jacob, *Robert Boyle and the English Revolution*, pp. 133–139. Jacob argues that Boyle, Barlow, and Pett intended these works on toleration "to serve commerce and industry as well as clergy and gentry" (p. 137). As Michael Hunter correctly points out, however, the stress on secular considerations that is particularly apparent in Pett's work is alien to Boyle's thought. Boyle may well have agreed that toleration was good for trade and public peace, but his primary concern was to do what was spiritually correct; see Hunter's Introduction to *Robert Boyle by Himself and His Friends*, p. lxix.

conscience was spurred by the fact that he himself might become persuaded of the truth of some doctrine not admitted by the established
church. Somewhat later he would write that

> in matters of Religion, the Authority of the Church with the general
> consent of Learned men, may sway with me, as long as I have no
> cogent Reason to be of another Opinion; but if I can light upon any
> demonstrative proof for a differing Opinion, I would follow my
> own private judgment against the general consent. As if I be to set
> my watch in weather that hides the sun from us, I am content to set
> it by the Town clock, or that which belongs to the Church holds,
> but if I have the opportunity of consulting a good sun-dial I shal not
> scruple, in case of difference (for some hours after) to prefer the
> information of my own watch to that of the Town clock, & as many
> other clocks & watches as were set by it, and agree with it.[33]

Whatever considerations may have contributed to Boyle's views on
liberty of conscience, those views were greatly influenced by his belief
that human beings cannot comprehend fully all the revelations contained
in scripture. His views on the limits of human understanding, especially

---

33. Boyle Papers, vol. 5, fol. 91. The hand is that of Robin Bacon, Boyle's chief
amanuensis after the mid-1670s, so the passage was either written or
transcribed after that date. A strikingly similar passage occurs in *Reason
and Religion*, published in 1675 (*Works*, vol. 4, pp. 155–156). The question of Boyle's conformity has been raised by Maddison (*Life*, pp. 139–
140), primarily on the grounds of Boyle's refusal to serve as president of
the Royal Society in 1680; in declining the position, Boyle mentioned his
reluctance to take the oaths required for the position (the Test Acts of 1673
and 1678). Generally, however, scholars have assumed that the picture of
orthodoxy described in memoirs written by John Evelyn and in Gilbert
Burnet's funeral sermon (both of which are now conveniently available for
the first time in *Robert Boyle by Himself and His Friends*) was accurate.
Evelyn noted that despite Boyle's "free thoughts" on religious matters
and discipline, in practice Boyle was "always conformable to the present
settlement" (British Library Add MS. 4229 fol. 59r; in *Robert Boyle by
Himself and His Friends*, p. 88). Burnet noted Boyle's financial support of
nonconformists, but added that Boyle was "constant to the Church; and
went to no separated Assemblies" (*Sermon Preached at the Funeral*, p. 29;
in *Robert Boyle by Himself and His Friends*, p. 50). In light of the most
recent scholarship, it seems likely that those who wrote memoirs of Boyle
thought it expedient to present him as more orthodox than he actually
was; see, for example, Michael Hunter, "How Boyle Became a Scientist"
("Appendix: Boyle and the Sects," pp. 86–92), which provides a valuable
corrective for the claim of James R. Jacob that Boyle's natural philosophy
was based on a "dialogue with the sects" (*Robert Boyle and the English
Revolution*, p. 87).

as expressed in his *Discourse of Things above Reason,* is a theme I explore fully in later chapters. Here, however, it is important to note that even in his early writings, Boyle revealed his views on the limits of human understanding and, even more importantly, indicated that the failure of those participating in doctrinal controversies to recognize those limits contributed to the divisions among Protestants. In his "Dayly Reflection," for example, he noted that

> surely he that shall impartially and unbyass'tly consider, how vast a Portion of Knowledge is yet Problematicke, how little of Truth we are groundedly and thoroughly satisfy'd our selves, and how much lesse of it we are able clearly and undeniably to demonstrate unto others, will think the Points not few in which 'twere very unreasonable to pass harsh Censure upon all Dissenters.[34]

It was because he believed human reason is incompetent to comprehend fully the content of scriptural revelations that Boyle deplored the dogmatic assertions of some of his contemporaries that they had attained a uniquely correct interpretation of certain disputed passages of scripture.

"Love of Truth" and "Love of Peace"

"A moderate and well-temper'd man," Boyle once noted, "in those controversies in Divinity where the Love of truth dos not engage him to spake, will let the Love of peace keep him silent."[35] In his initial reaction to the controversies over predestination, as expressed in *Some Motives and Incentives to the Love of God* (more commonly known as *Seraphic Love*), his love of peace led him to treat the differing interpretations as differences in emphasis rather than as substantive differences, clearly a conciliatory move, as I shall argue in chapter 3. Where the spread of Socinianism was concerned, however, Boyle's love of truth won out, at least initially. In his "Essay of the holy Scriptures," most likely written in the early 1650s and apparently an early version of his *Style of the Scriptures* (1661), he took exception to what he considered to be the doctrinal errors of the Socinians. Although his comments were at all times gentlemanly and detached in tone and quite mild compared with the heated polemics of his contemporaries, this direct and sustained attack on the Socinians was never published. I discuss Boyle's comments in chapter 2. Here what is important is to note that his suppression of such (for him) polemical writings was deliberate.

---

34. Boyle Papers, vol. 7, fol. 284; the passage may be found in Harwood, *Early Essays,* pp. xxxix and 228.
35. Boyle Papers, vol. 5, fol. 32v.

By the time Boyle prepared *Style* for publication, he had apparently reconsidered the consequences of including his comments about the Socinians, for no mention of them appears in *Style*. Exactly why he decided to suppress such passages is not clear, but there are several possible explanations, although no single explanation accommodates all of the facts easily. One is that by 1661, Boyle's reputation as a natural philosopher was well-established, and one of his primary concerns as a widely known and respected natural philosopher was to show not only that the corpuscular philosophy was not a threat to Christianity but that indeed it could be used to support the truths of the faith. With this goal in mind, perhaps Boyle feared that any participation in doctrinal controversies might lower his status as an eminent defender of Christianity as a whole. That Boyle saw himself as a naturalist writing in defense of Christianity in both the "Essay" and in *Style* is shown by the fact that a lengthy section of the "Essay" was devoted to a discussion of possibility of the resurrection of the body in corpuscular terms (although this discussion, too, was omitted from *Style*), as well as by his noting in the preface to *Style* that one of his purposes in publishing the work was to counteract the idea that a naturalist (and especially one who "explicates the phænomena of nature by atoms") is inclined to atheism, "or at least to an unconcernedness for any particular religion."[36] His discussion of the resurrection, incidentally, did eventually find its way into print (as *Physico-Theological Considerations concerning the Possibility of the Resurrection*, published along with *Reason and Religion* in 1675), whereas in none of his published works is there a sustained discussion of Socinianism.

Another possible explanation is that Boyle had experienced almost another full decade of the doctrinal controversies of mid-seventeenth century England between writing the "Essay" and the publication of *Style*. In February 1646, he wrote to Francis Tallents, a divine and fellow of Magdalene College, Cambridge, that

> few days pass here, that may not justly be accused of the brewing or broaching of some new opinion. Nay, some are so studiously changeling in that particular, they esteem an opinion as a diurnal, after a day or two scarce worth the keeping. If any man have lost his religion, let him repair to *London,* and I'll warrant him, he shall find it: I had almost said too, and if any man has a religion, let him but come hither now, and he shall go near to lose it. Pray God, it fare not with religion amongst those novelties, as it does sometimes with a great commander, when he is taken prisoner by a company

36. Boyle, *Style of the Scriptures, Works,* vol. 2, p. 253.

of common soldiers, who every one tugging to have him for himself,
at last pull him to pieces, and so each get a limb, but none enjoys
him whole.[37]

Perhaps he had decided not to give even the appearance of being one of
the "common soldiers." Of significance in this connection may be the
fact that during the two years between the 1660 Restoration and the
1662 Act of Uniformity (and *Style of the Scriptures* was published in
1661), any works contributing to doctrinal divisiveness might have been
perceived as lessening the possibility of the establishment of a comprehensive (or at least tolerant) post-Restoration Anglican church. During
these years, Boyle's "love of peace" may well have overcome his "love
of truth."

Another possible explanation, and one consistent with those suggested
here, is that by 1661 Boyle realized that the inclusion of any polemical
comments in *Style* might be inconsistent with his vision of the work as a
whole. As in the "Essay," he was concerned in *Style* to emphasize the
duty of lay Christians to study scripture. Yet one of the hazards of such
study was the possibility of differing (and incompatible) interpretations
of Christian doctrine. Indeed, this was one of the arguments of Roman
Catholics against the availability of scripture in the vernacular, and was
an argument Boyle had expressly noted in the "Essay."[38] For Boyle to
have included in *Style* any extended discussion of a doctrine he considered to be heretical (such as the anti-Trinitarianism of the Socinians)
would surely have undermined his purposes, one of which was to silence
"clamorous controversies" by arguing that scripture should be studied
as a whole, rather than as a collection of isolated passages, which
controversialists use as "loose stones, which they may more easily throw
at their adversaries, than built up into a structure."[39] Indeed, Boyle used
himself as an example in *Style*, noting that although "I neglect not those
clear passages or arguments, that may establish the doctrine of the
church I most adhere to; yet I am much less busied, and concerned to
collect those subtile glosses or inferences, that can but enable me to serve
one subdivision of Christians against another."[40]

Whatever his reasons for suppressing the passage concerning the
Socinians in the published version of *Style*, it was a conscious decision.

---

37. Boyle to Tallents, Feb. 1647, *Works*, vol. 1, p. xxxv. Tallents had tutored
    some of Boyle's brother Roger, Baron Broghill's, wife's relatives; Boyle
    most likely met him during a visit to Cambridge in December, 1645.
38. Boyle Papers, vol. 7, fols. 2, 5.
39. Boyle, *Style of the Scriptures*, *Works*, vol. 2, p. 275.
40. Boyle, *Style of the Scriptures*, *Works*, vol. 2, p. 276.

In the dedication to his brother, the Earl of Orrery, Boyle mentioned having shown Orrery "divers of these papers, with others (that I yet suppress)."[41] And in his preface, he noted that in the "essay of the scripture" of which his current work was a part, "divers things were interwoven . . . which were not so fit for publick view," and, as a result, Boyle had been forced "not only to dismember, but to mangle" the "Essay," by "cutting out with a pair of scissars" the parts he wished to omit.[42]

Further, the suppression of such passages was, or became, a matter of policy for Boyle, for none of his published works subsequent to *Seraphic Love* contains any sustained explicit discussion of doctrinal matters, except for his *Physico-theological Considerations about the Possibility of the Resurrection,* appended to *Considerations about the Reconcileableness of Reason and Religion* (1675).[43] Apparently he was occasionally tempted to include such discussions, but was encouraged to keep silent on doctrinal matters by at least one friend. In 1663, John Beale responded to Boyle's request that Beale read and comment on an essay Boyle had written, an essay in which Boyle had again attacked the Socinians. (It is interesting to speculate that this essay might have consisted of portions of his "Essay of the holy Scriptures" that were omitted from *Style of the Scriptures*.) It would be much better, Beale advised, if

41. Boyle, *Style of the Scriptures, Works,* vol. 2, p. 248.
42. Boyle, *Style of the Scriptures, Works,* vol. 2, p. 252.
43. *Possibility of the Resurrection* appears, on the surface, to enlist the corpuscular philosophy on behalf of Christian doctrine, and clearly this was one of Boyle's concerns. However, questions about the status of the soul between death and the resurrection of the body and the nature of the resurrected body were hotly debated in Boyle's day (see, for example, Norman T. Burns, *Christian Mortalism from Tyndale to Milton* [Cambridge: Harvard University Press, 1972]). In this case, Boyle was using the new science to support a particular interpretation of Christian doctrine. Specifically, he was arguing that it is physically possible (when God's omnipotence is taken into consideration) for the resurrected body to be based on the earthly body in such a way as to preserve physical personal identity (that is, in such a way that the resurrected body could be considered to be the exact same body as the physical one). Boyle was able to participate in this controversy without appearing contentious because he did not name those against whom he was arguing; see Davis, " 'Parcere nominibus'." Further, he maintained a detached, rational, and gentlemanly approach in an age noted for its heated polemics. See also my comments in note 45.

you do not (since you need not) engage yourself expressly against the interpretation of *Socinus*. ... Sir, in this I commend your prudence, that you keep the garb that is proper for a person of honour, and leave us choleric men of the lower region to answer to challenges, and to fight in duels.[44]

Apparently, Boyle took Beale's advice to heart. He himself did not fight in theological duels – at least not explicitly – and hence earned his reputation as being one who was above the heated polemics of seventeenth-century England.

Indeed, he came to wish that his contemporaries were not so willing to "answer to challenges, and to fight in duels." Not until the late 1670s and his involvement with John Howe's *Reconcileableness of God's Prescience of the Sins of Men, with the Wisdom and Sincerity of his Counsels, Exhortations, and whatsoever Means He uses to prevent them* (1677) and the ensuing controversy (which I discuss in chapter 3) did Boyle publicly involve himself in doctrinal controversies. Even then, Howe's *Reconcileableness* was a proposal of a *via media* between Arminianism and Calvinism, itself conciliatory. When it resulted in polemics rather than peace, Boyle went on to publish works specifically concerned with reason's limits – his *Discourse of Things above Reason* and its accompanying *Advices in judging of Things said to transcend Reason* (1681), as well as his *Reflections on a Theological Distinction*, appended to *The Christian Virtuoso* (1690). In these works, Boyle argued that it is inappropriate to make dogmatic claims concerning the truth of one or another interpretation of disputed passages of scripture when those passages contain revelations that are above (and at times even contrary

---

44. Beale to Boyle, 17 Oct. 1663, *Works*, vol. 6, pp. 342–343. Boyle referred briefly to Socinus in his 1674 *Reason and Religion* (*Works*, vol. 4, p. 172), and in fact quoted a passage from Socinus's second letter to Andreas Dudithius, which also occurs in the early "Essay" (Boyle Papers, vol. 7, fol. 46). It seems plausible that in an earlier version of *Reason and Religion* – perhaps the essay on which Beale commented – Boyle had included some of the suppressed passages from the "Essay"; on this, it is worth noting that *Possibility of the Resurrection* was published along with *Reason and Religion* and was itself related to the earlier "Essay." The letter from Beale to Boyle that I have quoted is missing from the collection of Boyle's papers and letters at the Royal Society in London, so it is impossible to determine to what extent, if at all, it was edited by Birch, who is known to have suppressed data that he considered to be trivial or contrary to the image of Boyle which he wished to portray; see Michael Hunter's Introduction to *Robert Boyle by Himself and his Friends*, pp. lxi-lxii.

to) human reason. In these same works, Boyle participated in controversies on reason's limits in the hope of putting an end to heated religious polemics, and he did so in a gentlemanly and restrained manner, intending to serve as an example to his contemporaries as to the proper way to discuss theological issues.[45]

Before turning to Howe's *Reconcileableness* and Boyle's ensuing works concerning reason's limits, it will be necessary to examine the controversies over the claim of the Socinians that human reason is to be the criterion against which differing interpretations of scripture should be measured and the controversies between the Arminians and the Calvinists over the correct interpretation of predestination, for these were the controversies that prompted his *Things above Reason*. Boyle's love of peace may have kept him almost totally silent on doctrinal matters in the 1660s and 1670s; his contemporaries, however, were busily "answering challenges and fighting duels." First, however, it will be helpful to survey briefly the question of the proper relationship of reason to revelation in the history of Christianity prior to Boyle's time, with an emphasis on the concepts that Boyle would use in *Things above Reason*.

45. Boyle's "gentlemanliness" may have reflected also a desire not to elicit polemical responses from adversaries; on this, see Michael Hunter's Introduction to *Robert Boyle Reconsidered*, p. 13. Hunter has also discussed the fact that Boyle's own doubts may have contributed to his reluctance to make dogmatic claims (ibid.). In discussing Boyle's rhetorical strategy, Edward B. Davis has stressed the significance of Boyle's desire to elicit the support of fellow Christians in the battle against atheism; see " 'Parcere nominibus'," esp. pp. 162–168. Each of these explanations is consistent with my account of Boyle's "gentlemanliness and restraint." Aspects of Boyle's gentility have been stressed in Steven Shapin, *A Social History of Truth: Civility and Science in Seventeenth-Century England* (Chicago: University of Chicago Press, 1994).

PART I

The Theological Context

1

Things above Reason: Medieval Context and Concepts

When confronted with the charge that the doctrine of Christ's Incarnation was irrational, the African Church Father Tertullian (c. 160–c. 220) cheerfully responded, "And the Son of God died; it is by all means to be believed, because it is absurd. And He was buried, and rose again; the fact is certain, because it is impossible."[1] It is likely that his claim was based on a passage in Aristotle's *Rhetoric*, where Aristotle explored a position that could be taken in response to the charge that certain beliefs are incredible: "We may argue that people could not have believed them if they had not been true or nearly true: even that they are the more likely to be true because they are incredible."[2]

Most likely, with his "I believe because it is absurd," Tertullian meant nothing more than that the objects of faith, even if incomprehensible to human understanding, are by virtue of their being revealed most certainly true. As Étienne Gilson observed in commenting on this passage, "Ever since the time of Athenagoras, theologians have been most anxious always to establish at least the *rational possibility* of the Christian faith."[3] This anxiety led to the claim that although some Christian doctrines may be above reason, none are contrary to reason, a claim that had become a commonplace expression by the seventeenth century.

1. Tertullian, *On the Flesh of Christ*, translated by Peter Holmes, in *The Ante-Nicene Fathers: Translations of the Writings of the Fathers down to A.D. 325*, edited by Alexander Roberts and James Donaldson, revised by A. Cleveland Cox, 10 vol., American reprint of the Edinburgh Edition (Grand Rapids, MI: Eerdmans, 1957), vol. 3, p. 525.
2. Aristotle, *Rhetoric*, translated by W. Rhys Roberts, in *The Basic Works of Aristotle*, edited by Richard McKeon (New York: Random House, 1941), Bk. II, Ch. 23, Sect. 21 (p. 1426). Robert M. Grant has noted this connection, although he mistakenly cited Sect. 22; see "Tertullian" in the *Encyclopedia of Philosophy*, edited by Paul Edwards, 8 vol. (New York: Macmillan and Free Press, 1969), vol. 8, p. 95.
3. Étienne Gilson, *History of Christian Philosophy in the Middle Ages* (New York: Random House, 1955), p. 45.

"Above but not contrary to reason," however, has meant different things to different Christian thinkers at different times and in different contexts. In this chapter, I explore some of those differences, providing, in the process, a partial listing of concepts inherent in the Christian tradition that Robert Boyle had on hand in his preparation of *A Discourse of Things above Reason* (1681).[4]

Irenaeus and Clement of Alexandria

Irenaeus (born c. 126), Bishop of Lyons at the end of the second century, contributed to the tradition of things above reason by arguing (against the Gnostics) that human knowledge has its limits, not only where the mysteries of Christianity are concerned, but also where the secrets of nature are concerned. It should not be surprising, he claimed, if we lack knowledge of God's mysteries,

> since many even of those things which lie at our very feet (I mean such as belong to this world, which we handle, and see, and are in close contact with) transcend our knowledge, so that even [knowledge of] these we must leave to God.

Among the secrets of nature not understood, Irenaeus cited the cause of the rising of the Nile, the out-of-season dwelling places of birds, the ebb and flow of the ocean, the causes of rain, lightning, thunder, and other weather phenomena, and the reasons for differences among waters, metals, and stones. "If, therefore, even with respect to creation," he concluded,

> there are some things [the knowledge of] which belongs only to God, and others which come within the range of our own knowledge, what ground is there for complaint, if, in regard to those things which we investigate in the Scriptures (which are throughout spiritual), we are able by the grace of God to explain some of them, while we must leave others in the hands of God.[5]

Or as Boyle would say some 1,500 years later in the seventeenth century,

---

4. It should be noted that I do not pretend that the very brief points I extract from the writings of the philosophers discussed in this chapter represent the thought of each as a whole; my goal is simply to provide a partial inventory of concepts inherent in the Christian tradition and available to Boyle as part of his heritage as a Christian philosopher as he struggled to form his own beliefs about the relationship between revelation and reason, emphasizing points that are relevant to his thought on "things above reason."

5. Irenaeus, *Against Heresies*, translated by Alexander Roberts and James Donaldson, in *Ante-Nicene Fathers*, vol. 1, p. 399.

one of the best, though least thought of Uses of Humane Reason, consists, *First,* in discovering how far its Powers can reach, and to what things they do not extend, and *then* in noblely attempting to surmount Difficulties that *are* superable, and wisely submitting itself to those that *are not* so.[6]

In Irenaeus's view human understanding is imperfect. Irenaeus attributed this deficiency to the fact that man, as a created being, is finite and limited, and often expressed the thought that humans are in a state of infancy, echoing the words of St. Paul in I Corinthians, chapter 3, verse 2 ("I have fed you with milk, and not with meat: for hitherto ye were not able *to bear it,* neither yet now are ye able").[7]

Not all early Christian apologists agreed with Irenaeus. Clement of Alexandria (born c. 150), had a different conception of the relationship between reason and faith. Arguing that philosophy is one of the works of God, it is, as such, good, and should be used in the service of Christianity. Although acknowledging that philosophy without Christianity is only partial truth, Clement claimed that the Christian, who already has access to the one whole truth of Christianity, should enlist philosophy as a tool for leading men to Christ, as well as an aid in inquiring into the meaning of faith after conversion. The true Gnosis is Christian, with faith blossoming into intellectual knowledge.[8]

These two trains of thought – the one that the human intellect is imperfect and incapable of understanding the mysteries of Christianity and that pretensions to an intellectual comprehension of the incomprehensible is the origin of heresies, and that the *Christian* intellect, improved by revelation and aided by the Holy Spirit, should enlist philosophy as an aid to understanding and as a weapon against heresies – were appealed to repeatedly thoughout medieval Christendom. Some thinkers stressed the one, and others the other. St. Thomas Aquinas was one who attempted to reconcile the two schools of thought.

Thomas Aquinas

The reintroduction of Aristotle's thought in the West in the late twelfth and early thirteenth centuries, and the ensuing concern that truths of

6. Boyle Papers, vol. 1, fol. 64.
7. Gilson, *History of Christian Philosophy in the Middle Ages,* p. 23 and p. 563 n. 67.
8. Gilson, *History of Christian Philosophy in the Middle Ages,* p. 33. See also Eric Osborn, *The Emergence of Christian Theology* (Cambridge: Cambridge University Press, 1993), pp. 272–274.

natural philosophy might trample upon the objects of faith, led to the remarkable effort of Thomas Aquinas (1225–1274) to reconcile reason and religion.[9] Despite the attraction of Aristotle's natural philosophy, some aspects of Aristotle's thought threatened certain of the revealed truths of Christianity. Aristotle had taught that the world is eternal, whereas scripture attests to its creation. Aristotle's world (especially as interpreted by the Latin Averroists) was ruled by necessity, whereas the Christian world was ruled by an omnipotent God.

Aquinas responded to this challenging situation by separating the objects of knowledge and those of faith. The truths of philosophy, Aquinas claimed, are based on natural reason alone, whereas the truths of faith are based on revelation; God, both the creator of human reason and the author of revelation, would not allow the two truths to conflict. Whatever God has revealed is most certainly true, whereas the truths of natural reason, being limited to the sensible world, are only probably true. Natural reason is perfectly adequate for an understanding of its proper objects, but it is an inadequate tool to steer the human soul to its final end. Knowledge of God, an infinite being, cannot be attained by finite human reason.

Despite its shortcomings, reason, in Aquinas's view, does have a role to play in theology. Unaided reason can grasp certain theological truths, such as the truths that God exists and that there is only one God. Where revelation is involved in these truths, it serves, first, to confirm the knowledge attained by reason alone, and, second, it makes such truths known to individuals who might not have attained them on their own. Further, philosophical reason can be appealed to in order to refute heretical doctrines by showing that they are clearly in error, or at least that they have not been demonstrated to be true. Most importantly as far as this book is concerned, human reason can aid the Christian in understanding the objects of faith by showing *how* a particular object of faith can be true, thus raising theology to the level of *scientia,* or demonstrative knowledge.[10]

9.　For the discussion of Aquinas, I have relied heavily on Étienne Gilson, *The Christian Philosophy of St. Thomas Aquinas,* translated by I.T. Eschmann (New York: Random House, 1956); St. Thomas Aquinas, *Faith, Reason and Theology: Questions I-IV of his Commentary on the "De Trinitate" of Boethius,* translated with introduction and notes by Armand Maurer (Toronto: Pontifical Institute of Mediaeval Studies, 1987); and Ralph McInerny, *Aquinas Against the Averroists: On There Being Only One Intellect* (West Lafayette, IN: Purdue University Press, 1993).

10.　Armand Maurer, Introduction to *Faith, Reason and Theology,* pp. xii-xiii.

What reason unaided by revelation cannot do, Aquinas claimed, is reach a knowledge of the truths necessary for salvation. These are the objects of faith, accepted on the basis of God's word. They are above human reason; God *could* have made them accessible to human reason but did not. And unaided human reason, being based on sensibles, certainly cannot contradict the revelation of God concerning truths about insensible objects. If God has revealed that he is three-in-one, then he is; if human reason shows that it is impossible for three to be one, then human reason is correct – *but only within its proper sphere* – which does not include a knowledge of God's essence.

Double-truth and the Law of Noncontradiction

At the same time that Aquinas was building his comprehensive synthesis of Aristotelianism and Christianity, certain contemporaries were handling the conflict between faith and reason in a different way, adhering to a doctrine of "double-truth." Tradition has associated this doctrine with Siger of Brabant, master of arts at the University of Paris during the third quarter of the thirteenth century, and although scholars have been unable to find the doctrine in Siger's extant writings, it does seem clear that it was the position taken by a number of Latin Averroists at the time.[11]

In *On There Being Only One Intellect* (1270), Aquinas himself summarized the doctrine, explaining that there are those who say,

> "Through reason I conclude necessarily that intellect is numerically one, but I firmly hold the opposite by faith." Therefore [they think] that faith is of things whose contrary can be necessarily concluded; since the only thing that can be necessarily concluded is a necessary truth whose opposite is false and impossible, it follows from this statement that faith is of the false and impossible, which not even God can bring about and the ears of the faithful cannot bear. [They do] not lack the high temerity to presume to discuss what does not pertain to philosophy but is purely of faith, such that the soul suffers from the fire of hell, and to say that the teaching of the doctors concerning these things should be reprobated. By equal right, one could dispute concerning the Trinity, the Incarnation, and the like, concerning which he speaks only out of blindness.[12]

---

11. McInerny, Introduction to *Aquinas Against the Averroists*, pp. 1–15, and interpretive essay "Double-Truth Theory," pp. 212–214. See also Gilson, *History of Christian Philosophy in the Middle Ages*, pp. 387–402.

12. St. Thomas Aquinas, *On There Being One Intellect*, in McInerney, *Aquinas*

A person who holds that it is necessarily true (because demonstrated by human reason) that there is only one intellect *and* that it is necessarily true (because revealed by God) that there are many intellects is clearly a person who holds contradictory beliefs. Aquinas could not and would not countenance this possibility; it amounted, in his opinion, to the claim that an article of faith was "false and impossible." Any such contradiction between reason and faith must be only apparent. Although God, the author of both reason and faith, is omnipotent, even he is bound by the law of noncontradiction. As Aquinas put it, "whatever implies contradiction does not come within the scope of divine omnipotence," because whatever implies a contradiction is impossible.[13] Whatever God has revealed is most certainly true. Hence, if natural reason reaches some conclusion contradictory to revealed truth, there must be some problem either with the conclusion reached by unaided natural reason or with the subject matter to which it is being applied. God might have revealed many truths above reason, but none contrary to reason.

Aquinas was working within a long line of theologians who had almost unanimously agreed that God's omnipotence is limited by the law of noncontradiction.[14] The double-truth doctrine, however, raised the question of whether one could coherently hold that two contradictory statements are both true. The Condemnations of 1270 and 1277 made it clear that this was not a position acceptable to the Church; before listing the condemned propositions, Étienne Tempier had warned

---

against the Averroists, p. 143. The issue involved the claim of the Averroists that Aristotle had proven that there is only one universal intellect. This is of course contrary to the Christian view that each individual has an intellect that continues to exist after the death of the body because it entails a loss of personal identity in the afterlife.

13. Thomas Aquinas, *Summa Theologica I*, Question 25, Article 3, in *Basic Writings of Saint Thomas Aquinas*, 2 vols., edited by Anton C. Pegis (Random House: New York, 1944), vol. 1, p. 263.

14. Peter Damian (1007–1072) was an exception; Damian argued that God, the author of the laws of logic, could change the laws at will. See Margaret J. Osler, *Divine Will and the Mechanical Philosophy: Gassendi and Descartes on Contingency and Necessity in the Created World* (Cambridge: Cambridge University Press, 1994), p. 19; Francis Oakley, *Omnipotence, Covenant, and Order: An Excursion in the History of Ideas from Abelard to Leibniz* (Ithaca, NY: Cornell University Press, 1984), pp. 42–44; Gordon Leff, *Medieval Thought: St. Augustine to Ockham* (London: The Merlin Press, 1959), pp. 96–97.

that those who had advocated the double-truth theory could not use it to defend any condemned proposition.[15] The Condemnations, directed primarily against those who attributed necessity to nature and by implication set boundaries to God's omnipotence, resulted in an emphasis on the absolute contingency of the created world and God's divine freedom to intervene in the creation, an emphasis that led logically to the view that an *a priori* knowledge of nature is impossible.[16] Despite this emphasis on God's omnipotence, however, his powers remained limited by the law of noncontradiction. William of Ockham (c. 1285–1349), for example, one of the foremost nominalist thinkers in the immediate

15. Gilson, *History of Christian Philosophy in the Middle Ages,* p. 406. The question of the double-truth doctrine would reappear from time to time. See, for example, the controversies surrounding Pietro Pomponazzi's *Tractatus de immortalitate animae* (1516) discussed by Charles H. Lohr, "Metaphysics," in *The Cambridge History of Renaissance Philosophy,* edited by Charles B. Schmitt, Quentin Skinner, Eckhard Kessler, and Jill Kraye (Cambridge: Cambridge University Press, 1988), pp. 537–648, esp. pp. 602–604; Brian P. Copenhaver and Charles B. Schmitt, *Renaissance Philosophy* (New York: Oxford University Press, 1992), pp. 104–112; Richard H. Popkin, "Theories of Knowledge," in the same volume, pp. 668–684, esp. pp. 670–671. As Popkin points out, if individuals accused of adhering to a double-truth doctrine were sincere in their protestations that the proposition of faith must be true (and hence that the proposition of philosophy must be false), the position then becomes one in which the limits of human reason are emphasized rather than reason's competence.

16. The Condemnations reflected the perceptions of conservative theologians that Christianity was being polluted with alien elements (that is, the attempted synthesis of Aristotelian philosophy, especially as interpreted by the Averroists, with Christian revelation). For the context and aftermath of the Condemnations, see Edward Grant, "The Condemnation of 1277, God's Absolute Power, and Physical Thought in the Late Middle Ages," *Viator* 10 (1979), pp. 211–244; idem, *Physical Science in the Middle Ages* (New York: John Wiley and Sons, 1971; paperback edition, Cambridge: Cambridge University Press, 1977); idem, "Science and Theology in the Middle Ages," in *God and Nature: Historical Essays on the Encounter between Christianity and Science,* edited by David C. Lindberg and Ronald L. Numbers (Berkeley: University of California Press, 1986), pp. 49–75; Osler, *Divine Will and the Mechanical Philosophy,* pp. 15–35; Leff, *Medieval Thought,* pp. 224–251; Gilson, *History of Christian Philosophy in the Middle Ages,* pp. 387–410. I discuss the implications of the post-Condemnations emphasis on God's absolute power for Boyle's thought in chapter 8.

post-Condemnations period, emphasized that even God could not bring about that which is logically impossible.[17]

A little over a century later, however, Nicholas of Cusa (1401–1464) questioned the validity of the law of noncontradiction, a step that surely suggests the most stringent restriction possible on the ability of human reason to grasp truth. According to Cusa, finite human reason is incapable of grasping the totality of knowledge held in the infinite mind of God; each glimpse of that truth is only partial. Cusa insisted that there is, in God, a "coincidence of opposites," a total knowledge of all truths, some of which are in reality, from the human perspective, contradictory. The insistence of the Aristotelians on the law of noncontradiction made a mystical ascent to an intuitive knowledge of God impossible, and should be abandoned. The truly wise man, Cusa claimed, is the man who knows reason's limits.[18] Cusa put this claim to work in a context similar to the one that was of concern to Robert Boyle 200 years later. Constantinople had just fallen into Muslim hands when Cusa wrote *On the Peace of Faith* (1453). While Rome was calling for a new crusade, Cusa tried to reconcile the Islamic and Christian religions (indeed, all religions) by arguing that each religion represents but one aspect of the total truth. Although the Christian religion was clearly superior, divine truths may be found in others as well.[19] Boyle would be most directly concerned with reconciling disagreements among Christians, but he used essentially the same concepts to attempt to reconcile quarreling Christians that Cusa had used to reconcile Christians and non-Christians. Boyle's claim that two truths that are genuinely contradictory from the viewpoint of finite human reason must be reconcilable in God's infinite reason is reminiscent of Cusa's thought, as is Boyle's claim that various sects of Christians might each possess one aspect of the totality of religious truth that is known to God alone.[20]

17. Osler, *Divine Will and the Mechanical Philosophy*, pp. 29–30.
18. See Armand A. Maurer, "Nicholas of Cusa," in *The Encyclopedia of Philosophy*, vol. 5, pp. 496–498. Also see Jasper Hopkins, *Nicholas of Cusa on Learned Ignorance: A Translation and an Appraisal of De Docta Ignorantia* (Minneapolis, Arthur J. Banning Press, 1981), pp. 1–43, as well as Gilson, *History of Christian Philosophy in the Middle Ages*, pp. 534–540.
19. *Nicholas of Cusa on Interreligious Harmony: Text, Concordance and Translation of De Pace Fidei*, edited by James E. Biechler and H. Lawrence Bond (Lewiston, NY: E. Mellen Press, 1990).
20. See chapter 4 for Boyle's view that God in his infinite wisdom perceives the harmony between truths that seem to human beings to be contradictory. See the discussion of *Seraphic Love* in chapter 3 for Boyle's view that

Lorenzo Valla

Lorenzo Valla (1407–1467), the Italian humanist best known for having exposed the Donation of Constantine as a forgery and thus undermining claims of papal supremacy in secular matters, was another thinker whose handling of a longstanding problem – the possibility of reconciling God's prescience with the free will of human agents – would later be reflected in Boyle's thought.[21] In his *Dialogue on Free Will*, Valla affirmed God's foreknowledge of future contingents, and declared such foreknowledge to be consistent with the free will of human agents. In affirming God's will concerning election and reprobation, however, he refused to address the question of whether God's omnipotence is consistent with the free will of humans beings. The *Dialogue* ends with the affirmation that God's decrees of election and reprobation certainly seem to be inconsistent with the freedom of the human will. Nevertheless, he affirms both, exclaiming (in the words of St. Paul), "O the depth of the riches both of the wisdom and knowledge of God! how unsearchable are his judgments, and his ways past finding out!"[22] Even the angels, "who always see the face of God," do not understand how God's decrees are consistent with the free will of human beings.[23] Two centuries later, Boyle would affirm both God's prescience, which must be true because of God's perfect omniscience, and the free will of human beings, the truth of which is attested to by experience, and then declare the two truths

---

theological disagreements might be the result of different Christian sects perceiving different aspects of a total truth. In *Seraphic Love*, Boyle included some non-Christians in this argument. In his unpublished "De Diversitate Religionum," however, Boyle argued for the unique truth of Christianity (Boyle Papers, vol. 6, fols. 279–291). For the future publication of this essay, see my Introduction, note 12.

21. For a history of attempts to resolve the problem, see William Lane Craig, *The Problem of Divine Foreknowledge and Future Contingents from Aristotle to Suarez* (Leiden: E. J. Brill, 1988).

22. Lorenzo Valla, *Dialogue on Free Will*, translated by Charles Edward Trinkaus, Jr., in *The Renaissance Philosophy of Man*, edited by Ernst Cassirer, Paul Oskar Kristeller, and John Herman Randall, Jr. (Chicago: University of Chicago Press, 1948), p. 176; Valla was quoting Romans, chapter 2, verse 33.

23. Valla, *Dialogue on Free Will*, p. 178. For a slightly different interpretation, see Antonio Poppi, "Fate, Fortune, Providence and Divine Freedom," in *The Cambridge History of Renaissance Philosophy*, pp. 641–667, esp. pp. 648–650.

inconsistent from the perspective of finite human understanding, but consistent from the perspective of God's infinite knowledge.

Two Approaches Summarized

In this brief survey of the sixteen centuries of Christian tradition prior to Boyle's *Things above Reason* (1681), I have sought to illustrate each of the concepts that played a significant role in the controversies about things above reason that took place in the three decades prior to Boyle's publication of his own views, not to provide a comprehensive account of any given philosopher's thought. Generally speaking, the attempts to resolve the question of the relationship of reason to faith fell into two categories, and the seventeenth-century controversialists with whom Boyle was involved mirrored these two categories as they had developed throughout Christianity's history.

There were, first, those who emphasized reason's competence to tackle theological questions. Although those who emphasized reason's role were quick to acknowledge that reason alone could never have *discovered* the mysteries of Christianity – revelation was needed for that – reason was considered competent to aid the believer in *understanding* the content of what has been revealed. The believer might not be able to understand fully each and every article of faith; after all, a mystery would not remain a mystery if it were understood completely, and if the believer were able to grasp completely all that God has revealed, little or no room would be left for faith. Nevertheless, the emphasis was on using the believer's God-given gift of reason to try to understand as much as possible. After the Christian had come to believe *that* something was true simply because God had revealed it, the believer could analyze that truth philosophically in an attempt to determine *how* it could be true. Further, reason was seen as playing an important role in convincing atheists of the truth of Christianity by showing that the mysteries of Christianity are rationally possible, as well as in confuting heresies by enlisting reason on the side of orthodoxy. In short, reason as such was perceived as both competent and useful. In this brief survey, I have used Clement of Alexandria and Thomas Aquinas to illustrate this position.

Second, there were those who emphasized reason's limits, reason's *inability* grasp the content of the revealed mysteries of Christianity. In this view, human reason was *not* considered competent to understand the content of revelation – at least not in its present state. Some of those who held this view of reason emphasized the vast difference (indeed, the incommensurability) between finite human reason and the infinite wisdom of God. Others believed that human reason as God had created it had been a great gift, but that Adam, in placing his desire for wisdom above his obligation to obey God, had forfeited his own and his posterity's full enjoyment of the faculties God had originally given the human products of his creation: both human reason and human will had been debased and corrupted by Adam's fall. Nicholas of Cusa, Lorenzo Valla, and the nominalists emphasized the limits of human reason, as had Tertullian and Irenaeus centuries earlier, some individuals going so far as to suggest that either God is not bound by the law of noncontradiction, or that if he is so bound, human reason is incapable of grasping the consistency of all that God has revealed in scripture and in nature.

Anglicans and Puritans

Both of these traditions had found their way into the England in which Boyle matured and formed his own views. While everyone claimed to possess "right reason," or reason seasoned by and submissive to revelation, different groups had different conceptions of "right reason."[24]

24. Despite a number of studies on "right reason" in seventeenth-century England, the variety of uses to which claims of right reason were put makes generalizations (mine included) suspect. A number of scholars have argued that there was a dramatic shift from the private illuminist epistemologies of the Civil War years to a reliance on natural reason occasioned by the relatively widespread acceptance of the mechanical philosophy in the years following the Restoration. See, for example, P.M. Rattansi, "Paracelsus and the Puritan Revolution," *Ambix* 5 (1963), pp. 24–32; Christopher Hill, " 'Reason' and 'Reasonableness' in Seventeenth-Century England," *British Journal of Sociology* 20 (1969), pp. 235–252; Robert Hoopes, *Right Reason in the English Renaissance* (Cambridge: Harvard University Press, 1962). Lotte Mulligan has challenged that assessment, arguing that there was no such sharp discontinuity (" 'Reason', 'Right Reason', and 'Revelation'," in *Occult and Scientific Mentalities in the Renaissance,* edited by Brian Vickers [Cambridge: Cambridge University Press, 1984], pp. 375–401). In this, I think she is essentially correct, although it does seem that she has elided rather crucial ways in which various post-Restoration Anglicans differed in their conceptions as to exactly what right reason meant. Further, in comparing the illuminist epistemologies of the Civil War years with the right reason advocated by post-Restoration Anglicans, both Mulligan and other scholars writing on right reason have failed to take

Oversimplification is hazardous, but generally speaking the Puritans conceived of human reason as depraved by Adam's fall. Although they believed reason to be competent within its own sphere of natural knowledge, they claimed that reason is incompetent to comprehend the content of revelation. After the Restoration, this conception of reason continued to be particularly influential among the nonconformists. This is not to say that the Puritans and nonconformists did not *use* human reason in theological matters; indeed, many of them considered secular knowledge (of languages, of rhetoric, and so on) to be one of the most powerful weapons against perceived heresies. John Owen's *Vindiciæ Evangelicæ*, discussed in chapter 3, provides ample evidence of that. However, when compared with their Anglican countemporaries, the Puritans and nonconformists took refuge sooner and more frequently than did the Anglicans in the claim that human reason is incompetent to judge the content of scripture.[25] At the same time, they (and most particularly high Calvinist nonconformists such as Owen) did not hesitate to claim that they and they alone, aided by God's grace, held the true doctrines of Christianity, having correctly interpreted that which they themselves claimed to be incomprehensible, leading at least one fellow nonconformist to complain that "it were much to be wished that men had never gone about to explain those things which were mysterious, and to give an account of

into account views advocated by various nonconformists. A study concentrating on the differing *applications* of right reason made by a wide variety of groups of English *Christians* after the Restoration would reveal far less homogeneity than her study would lead one to expect. "Reason seasoned by revelation" meant radically different things to different individuals and groups of individuals in post-Restoration England, hence the claim of Jeremy Taylor that "reason is such a box of quicksilver that it abides no where; it dwells in no settled mansion . . . [and] it looks to me otherwise than to you" (Jeremy Taylor, *Ductor Dubitantium or the Rule of Conscience in all her generall measures* [London, 1660], vol. 1, p. 231, cited in Mulligan, "'Reason'," p. 394). Mulligan points out that Taylor's complaint was against (faulty) natural reason rather than right reason. Despite this, I believe that Taylor's comment accurately depicts the variety of appeals to right reason.

25. The best study of the Puritan tradition where reason is concerned is John Morgan, *Godly Learning: Puritan Attitudes towards Reason, Learning, and Education, 1560–1640* (Cambridge: Cambridge University Press, 1986), especially chapters 3 and 4. See also Isabel Rivers, *Reason, Grace, and Sentiment: A Study in the Language of Religion and Ethics in England, 1660–1780*, vol. 1, *Whichcote to Wesley* (Cambridge: Cambridge University Press, 1991), chapter 3.

those things which [they] themselves acknowledg'd incomprehensible."[26]

The Anglican latitudinarians, in contrast to the Puritans, tended to emphasize the competence of human reason.[27] Influenced by the Cambridge Platonists, the post-Restoration latitudinarians emphasized, as had Aquinas, that God was the author of reason as well as revelation, and that although human reason is imperfect, it is competent to aid in understanding that which is believed on faith. Placing more emphasis on actual sin than original sin, the latitudinarians tended to believe that human reason in the Christian had been at least partially restored to the pristine nature forfeited by Adam's disobedience by virtue of Christ's obedience; for the Christian, human reason is "the Candle of the Lord." Further, God, the author of human reason, would not reveal something that contradicted his unique gift to man, the rationality that distinguishes human beings from animals.[28]

It is important to understand that both Anglicans and Puritans were quick to claim that some scriptural revelations are above human reason but not contrary to it. Most commentators have taken this common assertion at face value, but in fact different thinkers meant totally different things by it, depending on the context in which it was used. It could mean that scripture contains revelations that, although not discoverable by unaided human reason, could be understood once revealed. Or it could mean (as it did to certain of the nonconformists and to Boyle) that although no revelation *truly* violates the law of noncontradiction if viewed from the perspective of omniscience in which all truths are harmonious, a revelation could indeed *seem* to be contradictory to

26. Benjamin Calamy, *A Sermon Preached before the Right Honorable the Lord Mayor and the Court of Aldermen at Guild-Hall Chappel upon the 13th of July, 1673* (London, 1673), p. 15. The point of Calamy's sermon was that Christians should live moral lives, not engage in controversy.

27. See C.A. Patrides, Introduction to *The Cambridge Platonists*, edited by C.A. Patrides (Cambridge: Harvard University Press, 1970), pp. 9ff; Gerald R. Cragg, Introduction to *The Cambridge Platonists*, edited by Gerald R. Cragg (New York: Oxford University Press, 1968), pp. 16–20; John Spurr, *The Restoration Church of England, 1646–1689* (New Haven, CT: Yale University Press, 1991), esp. pp. 250–257, 270–272; and Rivers, *Reason, Grace, and Sentiment*, chapter 2.

28. Ernst Cassirer, *The Platonic Renaissance in England*, translated by James P. Pettegrove (Austin: University of Texas Press, 1953), pp. 38–41. C.A. Patrides notes that with the exception of Henry More, the Cambridge Platonists evinced an "unwillingness to so much as mention 'original sin' " (*The Cambridge Platonists*, p. 38).

human beings, and in fact *does* violate the law of noncontradiction from the point of view of finite and corrupted human understanding.[29] The claim could be that nothing in scripture is contrary to (God's) reason, or that nothing in scripture is contrary to (human) reason. The significance of this distinction has not been recognized by Boyle scholars, with the result that he has been portrayed as placing more emphasis on the rationality of Christianity than in fact he did.[30]

Boyle was situated firmly in the tradition of Tertullian, Irenaeus, the Ockhamists, Nicholas of Cusa, and Lorenzo Valla. Like Tertullian, he believed that some of God's revelations appear impossible to finite human understanding. Like Irenaeus, he was not surprised to find mysteries in revelation when he considered how many of nature's secrets are impenetrable by human understanding. Like the Ockhamists, he emphasized God's infinite power and will; indeed, the basis of Boyle's views on reason's limit was his belief that God had freely and deliberately chosen to limit the understanding of the human products of his creation. Like Nicholas of Cusa, he believed that human beings can grasp only aspects of God's total revelation, and like Cusa, he saw the relevance of this belief to the possibility of reconciling different and opposing theological views. And along the same lines as Valla, Boyle asserted that it is true both that God has foreknowledge of future contingents, and that human beings have free will, and that these are (from the perspective of finite human reason) truly contradictory propositions.

Insofar as his views on *natural religion* were concerned, Boyle had much in common with Aquinas: Boyle believed that *some* revelations are ascertainable by reason unaided by revelation and that revelation simply *confirms* those truths for those who have already grasped them. But in the long run, Boyle's emphasis was on the absolute incompetence of human reason to comprehend the more difficult of God's revelations.

While Boyle was developing his views on the question of the proper relationship of reason to revelation, Socinian ideas were spreading rapidly in England. The Socinians were a sect of radical reformers who insisted that difficult or apparently conflicting passages of scripture be interpreted in such a way as to be consistent with *human* understanding, and the spread of their views precipitated a prolonged controversy in England over the question of the role of reason in religious matters. As this controversy unfolded during the 1650s, 1660s, and 1670s, a number of Boyle's contemporaries argued within the long tradition of Christian

29. Although Boyle never denied that human reason was corrupted as a result of Adam's fall, his emphasis was on the finiteness of human reason.
30. I discuss the implications of this misplaced emphasis in the Conclusion.

thinkers who had emphasized the limits of human reason, who invoked the arguments concerning reason's limits made by their medieval predecessors, and who utilized each of the concepts that Boyle would weave together into a coherent whole in *Things above Reason*. It is to the arguments of the Socinians and to the responses of Boyle's contemporaries to Socinian arguments that I now turn.

2

The Threat of Socinianism

The Protestant insistence that scripture alone was to be the rule of faith –
and the accompanying claim that the laity was competent to interpret
scripture – brought with it the inevitable consequence of the proliferation of alternative and conflicting interpretations of scripture, and hence
the proliferation of sects. In this chapter, I examine the claims of one of
those sects, the Socinians, with an emphasis on their claim that there is
nothing in scripture that contradicts human reason. I also examine
some of the responses to Socinianism published in England prior to the
publication of Boyle's *Things above Reason* (1681), for it is in these
responses to Socinian claims that the rough outlines of the themes
developed in *Things above Reason* are to be found.

The Protestant Background

The claim that scripture contains doctrines that are above reason, but
none contrary to reason, was one of the battle cries of the Reformation
in England. William Chillingworth (1602–1644), for example, stressed
that each individual must use his God-given reason to make a conscious
choice between Roman Catholicism and Protestantism. In advocating
the study of scripture alone as a rule of faith, Chillingworth observed
that although nothing in *scripture* is contrary to reason, some doctrines
of *Roman Catholicism* require a person to believe things (such as Transubstantiation) that are impossible. As Chillingworth expressed this view
in his very influential *The Religion of Protestants a Safe Way to Salvation* (1638),

> following your [Roman Catholic] Church, I must hold many things
> not only above reason, but against it, if any thing be against it;
> whereas, following the scripture, I shall believe many mysteries, but
> no impossibilities; many things above reason, but nothing against it;
> many things, which had they not been revealed, reason could never
> have discovered, but nothing which by true reason may be confuted;

42

many things, which reason cannot comprehend how they can be, but nothing which reason can comprehend that it cannot be. Nay I shall believe nothing which reason will not convince [me] that I ought to believe it: for reason will convince any man, unless he be of a perverse mind, that the scripture is the word of God: and then no reason can be greater than this, God says so, therefore it is true.[1]

This passage from Chillingworth is interestingly ambiguous. On the surface it seems unproblematic, for in claiming that "reason will convince any man, . . . that the scripture is the word of God," Chillingworth could be interpreted as claiming that it is rational to believe anything that God has revealed in scripture (along with the implied claim that the doctrine of Transubstantiation is to be found nowhere in scripture). Chillingworth made it clear, however, that he was rejecting the Roman Catholic *interpretation* of a passage of scripture on the grounds that the interpretation entails impossibilities and hence the doctrine of Transubstantiation "by true reason may be confuted." Specifically he was denying that Matthew 26, verse 26 ("this is my body") should be interpreted literally.[2]

The Socinians were a group of radical reformers who carried this argument to its logical conclusion, claiming that *all* doctrines that "by true reason may be confuted" should be rejected, Protestant as well as Roman Catholic.

Early Socinianism

Radical reformers were Protestants (often lay persons) who, for various reasons, believed that the Reformation had failed to recapture the doctrinal purity and spirit of the early Christian church. Although there was often a considerable overlap in the beliefs of these groups, they can be

---

1. William Chillingworth, *The Religion of Protestants a Safe Way to Salvation; or, An Answer to a Booke entituled Mercy and Truth; or, Charity Maintain'd by Catholiques, which Pretends to Prove the Contrary*, 3 vols., 10th ed. (Oxford: 1838; reprint New York: AMS Press, 1972), vol. 2, p. 413.
2. Chillingworth, *The Religion of Protestants*, vol. 2, p. 81. For an excellent discussion of Chillingworth's *Religion of Protestants* within the context of the rule of faith controversy, see Henry G. van Leeuwen, *The Problem of Certainty in English Thought, 1630–1690* (The Hague: Martinus Nijhoff). For Chillingworth's role in the Great Tew Circle, see Hugh Trevor-Roper, *Catholics, Anglicans and Puritans: Seventeenth Century Essays* (London: Secker and Warburg, 1987), pp. 166–230.

roughly divided into Anabaptists, spiritualists, and rationalists.[3] The Socinians were the most significant and influential group of rationalists, a group that began with the teachings of Laelius Socinus (1525–1562) and his nephew, Faustus Socinus (1539–1604). Faustus Socinus's fast-growing fame brought him to the attention of a like-minded group of reformers living in Raków, Poland, where he moved in 1579; hence the group was sometimes known as the "Polish Brethren."

Laelius Socinus was a layman, a Venetian who abandoned Roman Catholicism, embraced psychopannychism,[4] and, influenced by Michael Servetus, denied the Nicene doctrine of the Trinity. When forced (in 1555, by Calvin and other reformers) to affirm his orthodoxy, he did so equivocally, asserting his right to inquire into and dispute about doctrinal matters as a means of enhancing his understanding.[5]

Faustus Socinus, in *De sacrae scripturae auctoritate* (begun in 1570, ultimately published in 1584), argued that although there may be some things in scripture that are above reason, there is nothing in scripture that is contrary to reason, and that salvation depends upon both a rational interpretation of scriptural doctrines and an ethical following of Biblical precepts. Socinus explained the obscurities and seeming contradictions in scripture by arguing that God had caused scripture to be written in such a way that honest enquirers would be able to interpret perplexing passages properly, whereas insincere enquirers would be unable to do so.[6] In *De Jesu Christo servatore* (finished in 1578, and circulated in manuscript form prior to its publication sixteen years later),

---

3. For a detailed history of the radical Reformation from 1516 through (approximately) 1580, see George Huntston Williams, *The Radical Reformation* (Philadelphia: Westminster Press, 1962); idem, editor and translator, *The Polish Brethren: Documentation of the History and Thought of Unitarianism in the Polish-Lithuanian Commonwealth and in the Diaspora, 1601–1685,* 2 vols. (Missoula, MT: Scholars Press, 1980). I rely heavily on Williams's detailed scholarship in my account of the development of Socinianism.

4. "Psychopannychism" is the belief that the soul is either dead ("thanetopsychism") or unconscious ("psychosomnolence") pending the resurrection of the body. Such a belief, of course, is inconsistent with the Roman Catholic doctrine of purgatory. See Norman T. Burns, *Christian Mortalism from Tyndale to Milton* (Cambridge: Harvard University Press, 1972).

5. Williams, *Radical Reformation*, pp. 567–570, 630–631. On Servetus (1511–1553), see idem, "Michael Servetus," in Paul Edwards, editor, *The Encyclopedia of Philosophy,* 8 vols. (New York: Macmillan and Free Press, 1967, reprint edition, 1972), vol. 7, pp. 419–420.

6. Williams, *Radical Reformation*, pp. 750–751.

he denied the essential deity of Christ, and claimed that when born Christ was merely a man (although a man miraculously conceived by the Virgin Mary in order to fulfill God's plan of salvation).[7] In developing his Christology, Socinus argued that Christ, like all human beings even before Adam's fall, was naturally mortal. Christ's significance, according to Socinus, was not his atoning death, but his resurrection, which served as a sign that God would eventually resurrect and grant eternal life to all believers.[8] His *Christianae religionis brevissima institutio,* on which he was working at the time of his death, became the basis of the first edition of the *Racovian Catechism* (published in 1605, the year following Socinus's death).[9]

7. Socinian anti-Trinitarianism consisted of the belief that Jesus Christ was a man miraculously conceived by Mary in fulfillment of Old Testament prophecies, and that despite his special status, his nature was fully human (at least prior to his resurrection, at which time he became the adopted son of God while remaining a distinct individual). This view should be distinguished from the belief of Arians that Jesus Christ shared in the divinity of God, was the first act of God's creation and was himself the creator of all else, but was not co-essential with God the Father. Both Socinians and Arians should be distinguished from another group of anti-Trinitarians, the Sabellians, who believed that the entire Trinity was incarnated in Jesus Christ, and that the fullness of the Godhead dwells in each believer by virtue of the union of the believer with the Holy Spirit. Arius's views were condemned at the Council of Nicea in 325, although they continued to be embraced by some of the faithful for another fifty years. Those views were finally defeated by the Athanasians, whose arguments for the co-essentiality and co-eternality of the Father and Son were accepted by the Council of Constantinople (381), at which the decision of 325 was reaffirmed and expressed more fully in the Athanasian Creed. Sabellius taught, during the latter half of the third century, that there is one God who assumes three offices rather than one God in three persons. For the details of creeds, I have relied on Philip Schaff, *The Creeds of Christendom, with a History and Critical Notes,* 3 vols. (New York: Harper, 1877).

8. Williams, *Radical Reformation,* pp. 749–756. Socinus's Christology was an attempt to avoid what he considered to be the contradictory interpretation of God's nature inherent in the traditional account of the atonement, in which God was believed to have accepted the punishment and death of his son as satisfaction for God's sense of justice; for an account of Socinus's perception of these contradictions, see especially pp. 754–756.

9. The authors of the original *Racovian Catechism* were Faustus Socinus, Peter Statorius, Jr., Valentine Smalcius, Jeromos Moskorzowski, and Johannes Volkelius (H. John McLachlan, *Socinianism in Seventeenth Century England* [Oxford: Oxford University Press, 1951], p. 36). See also fn 29.

The beliefs of the Socinians reached a more mature and developed formulation during the early seventeenth century, due primarily to the writings of such individuals as Johannes Crellius (1590–1633), Jonas Schlichting (1602–1661), and Joachim Stegmann (d. 1633). Like other Protestants, they pressed the idea that scripture alone is the rule of faith, and they soon gained a reputation for being extremely skilled at scriptural exegesis. However, they also explicitly insisted that scripture be interpreted rationally and that scriptural interpretations imply no contradictions. Although scripture was to be the rule of faith, reason was to be the judge of the proper interpretation of the rule. If an interpretation of a passage of scripture implied a contradiction, then an alternative interpretation of the passage in question must be sought by studying the passage more carefully in its original language and historical context. Because God is good, for example, those passages of scripture that could be interpreted as referring to absolute predestination should be interpreted in some other way, for the doctrine of absolute predestination, they argued, makes God the author of sin, which, because of his infinite goodness, he cannot be.[10]

In addition to being known for their particular doctrinal beliefs, their painstaking scriptural exegesis, and their insistence on the rational interpretation of scripture, the Socinians were also well-known for their advocacy of religious liberty; indeed, many seventeenth-century thinkers considered religious toleration to be a specifically Socinian doctrine.[11] In this, they were influenced by the "Pax dissidentium," formulated in Poland at a time when the throne was vacant (1573). This declaration of religious toleration was meant to ensure that whether the future king were Protestant or Roman Catholic, he would not persecute those who adhered to the other faith. When later developments resulted in the persecution of Protestants, Johannes Crellius wrote *Vindiciæ pro religionis libertate* (1632, published in "Eleutheropoli" [Amsterdam] in 1637), in which he argued that liberty of religion is lawful and right, and that the Roman Catholics, who had agreed to religious liberty in Poland when they had thought they might benefit, could not rescind that freedom when it suited them to do so.[12] In the long run, Crellius's effort was in vain: the "Pax dissidentium" was abandoned by King John II Casimir, who, in a series of Decrees of Banishment (1658–1661), re-

10. *Racovian Catechism*, pp. 142–146. Controversies over predestination provided the immediate context of Boyle's *Things above Reason* – see chapter 3. See fn 29 for a full citation.

11. McLachlan, *Socinianism in Seventeenth Century England*, p. 9 fn 1.

12. Williams, *Radical Reformation*, pp. 736–738, 749–756.

quired all Polish anti-Trinitarians either to convert to Catholicism or to leave Poland.[13]

In their emphasis on liberty of religion, as well as in their emphasis on the role of free will in the salvation process and the universality of God's offer of salvation, the Socinians had much in common with the Dutch Arminians.[14] Even before the Diaspora of 1658–1661, many Socinians had settled in the more congenial Netherlands, and it was primarily from there that Socinian writings found their way into England.

The "Englishing" of Socinianism

Although scholars have become increasingly aware of the significance of the Socinian influence in late seventeenth- and early eighteenth-century England, and especially its contribution to the rise of English deism, much work remains to be done, especially on Socinianism in England prior to the 1690s.[15] Although there were anti-Trinitarians in England during the second half of the sixteenth century (between 1548 and 1612 at least eight were burned at the stake),[16] beliefs that were specifically Socinian in nature were introduced primarily into England by means of Dutch immigrants and the importation of various editions of the *Racovian Catechism* and related writings. Indeed, the 1609 edition of the *Catechism* was dedicated to James I, apparently in the fruitless hope that

13. Williams, *The Polish Brethren*, vol. 2, pp. 490–496. For a more nuanced discussion of toleration and persecution in Poland during the Reformation, see Janusz Tazbir, "Poland," in *The Reformation in National Context*, edited by Bob Scribner, Roy Porter, and Mikuláš Teich (Cambridge: Cambridge University Press, 1994), pp. 168–180.

14. On the Dutch Arminians, see Rosalie Colie, *Light and Enlightenment: A Study of the Cambridge Platonists and the Dutch Arminians* (Cambridge: Cambridge University Press, 1957).

15. For general background information about Socinianism in England, see J. Hay Colligan, *The Arian Movement in England* (Manchester: At the University Press, 1913) and McLachlan, *Socinianism in Seventeenth-Century England*. For the relationship of Socinianism to English deism, see Zbigniew Ogonoswki, "Le 'Christianisme sans mystères' selon John Toland et les sociniens," *Archiwum historii filozofii i mysli spolecznej* 12 (1966), pp. 205–23; Robert E. Sullivan, *John Toland and the Deist Controversy* (Cambridge: Harvard University Press, 1982), esp. pp. 82–108; and J.A.I. Champion, *The Pillars of Priestcraft Shaken: The Church of England and its Enemies 1660–1730* (Cambridge: Cambridge University Press, 1991).

16. McLachlan, *Socinianism in Seventeenth-Century England*, p. 31.

such a dedication would facilitate the congenial reception of the work. By 1637, Socinianism was considered to be a significant enough threat to cause the Archbishop of Canterbury, William Laud, to procure a decree in the Star Chamber to regulate the printing trade and the importation of books from abroad. In 1640, Laud supported a number of laws that provided for the excommunication and trial of any importers, printers, or sellers of Socinian works, as well as the prohibition of any sermons advocating Socinianism and the forbidding of any university students (unless especially exempted) to read Socinian works. Further, all Socinian works found in unauthorized hands were to be burned. Although these laws were rejected by the House of Commons as being against the fundamental laws of the realm and as impinging on the rights of Parliament, concerned churchmen continued on a case-by-case basis to attempt to combat the spread of Socinian doctrines (as well as the spread of other doctrines perceived to be heretical).[17]

Francis Cheynell

Francis Cheynell was one such clergyman and the author of *The Rise, Growth, and Danger of Socinianisme; together with a plaine discovery of a desperate design of corrupting the Protestant Religion, whereby it appeares that the Religion which hath been so violently contended for (by the Archbishop of Canterbury and his adherents) is not the true Protestant Religion, but an Hotchpotch of Arminianism, Socinianism and Popery: It is likewise made evident, that the Atheists, Anabaptists, and Sectaries so much complained of, have been raised or encouraged by the doctrines of the Arminian, Socinian and Popish Party* (1643), a title that requires some explanation. The association of Socinianism with enthusiasts seems strange given the Socinian emphasis on the rationality of Christianity, but it can perhaps be explained by the fact that other anti-Trinitarians, the Sabellians, claimed that the entire Godhead dwells in each believer by virtue of the union of the Holy Spirit with the believer. The association with Arminianism was due to the emphasis of Arminians and Socinians alike on the role of the free will of human beings in the salvation process; this emphasis on free will, in turn, could be conflated with the alleged Pelagianism of the Roman Catholic communion, despite the fact that Pelagianism had been declared hereti-

---

17. McLachlan, *Socinianism in Seventeenth-Century England*, chapter 3. In a letter to Boyle on August 10, 1666, John Beale noted that Socinians had managed to elude Laud's censorship (*Works*, vol. 6, p. 413).

cal by the Council of Carthage in 411 (with subsequent denunciations in 417 and 418 that were due primarily to the efforts of Saint Augustine, who utterly rejected the Pelagian denial of the doctrine of original sin).[18] The reference to the "Archbishop of Canterbury," of course, was to William Laud, who was perceived by strict Calvinists to have imported Arminianism into the English church. The extent to which this perception might have been accurate is a matter of controversy.[19] Cheynell's accusing Laud of Socinianism is ironic, given Laud's attempts at censorship, and reveals the extent to which Socinianism and Arminianism were often conflated, although it must be admitted that in the body of the work, Cheynell revealed a more sophisticated understanding of Socinianism as a distinct set of beliefs than his title indicated.

Cheynell's effort was occasioned by the fact that a clandestine version of the *Racovian Catechism* was circulating in English. One Mr. Webberly, a "Batchelour" of Divinity and Fellow of Lincoln College, Oxford, had translated the catechism from Latin into English for (or so he claimed) "his own use," but, as Cheynell pointed out, if Webberly could translate Latin he could obviously read it, and hence had had no need for an English translation. Besides, on the frontispiece of the English version, Webberly had noted that he had translated the work "for the good of the nation," and he had also added a "To the Reader" section, surely strange signs of a work intended for personal use. Cheynell had originally hesitated to publish his rebuttal in English for fear of making Socinian views more widely available, but ultimately decided to do so on the grounds that Socinian ideas were already being debated, pro and con, in English.[20]

18. For the controversy between Augustine and Pelagius, see Henry Chadwick, *Augustine,* Past Masters (New York: Oxford University Press, 1986), pp. 107–117.

19. Nicholas Tyacke, for example, argues that Laud was indeed an Arminian in *Anti-Calvinists: The Rise of English Arminianism c. 1590–1640* (Oxford: Clarendon Press, 1987), esp. pp. 266–27. Peter White, on the other hand, argues that Laud was not; see *Predestination, Policy and Polemic: Conflict and Consensus in the English Church from the Reformation to the Civil War* (Cambridge: Cambridge University Press, 1992), esp. pp. 277–286.

20. Francis Cheynell, *The Rise, Growth, and Danger of Socinianisme; together with a plaine discovery of a desperate design of corrupting the Protestant Religion, whereby it appeares that the Religion which hath been so violently contended for (by the Archbishop of Canterbury and his adherents) is not the true Protestant Religion, but an Hotchpotch of Arminianism, Socinianism and Popery: It is likewise made evident, that the Atheists, Anabaptists, and Sectaries so much complained of, have been raised or*

John Dury

A further "Englishing" of Socinian ideas, one significantly more favorable than Cheynell's rebuttal of Socinianism, was John Dury's translation in 1646 of Johannes Crellius's *Vindiciæ pro religionis libertate* (1637). In his brief (and unpaginated) introduction, Dury argued that if Crellius's argument that liberty of religion is lawful and right holds true when the magistrate is a Roman Catholic, it is an even stronger argument when applied to a Protestant magistrate, because Protestants make no claims of infallibility, unlike the Roman Catholics. Although Dury was quick to point out that the Roman Catholic claim of infallibility is erroneous, he noted that it at least gives a Catholic magistrate grounds for denying liberty of religion; Protestants, on the other hand, have no grounds at all to deny religious liberty, for Protestants acknowledge that assemblies of divines may err (and have erred), and argue that "common people" are capable of searching the scriptures for truth.[21]

John Biddle

Perhaps emboldened and encouraged by responses to Socinian writings such as Dury's, or perhaps dismayed by rebuttals such as Cheynell's, or perhaps simply out of a desire to spread what he perceived to be the true Protestant doctrine, one Socinian, John Biddle, made Socinian thought widely available in English in the early 1650s, both by translating ex-

---

encouraged by the doctrines of the Arminian, Socinian and Popish Party (London, 1643), sig. A3r.

21.   Johannes Crellius [Junius Brutus, pseud.], *Vindiciæ pro religionis libertate*, translated by John Dury under the title of *A learned and exceedingly well-compiled Vindication of Liberty of Religion: Written by Junius Brutus in Latine, and Translated into English by N.Y. who desires, as much as in him is, to do good unto all men* (No Place [sic], 1646), Introduction (unpaginated). For Dury's irenicism, see Anthony Milton, " 'The Unchanged Peacemaker'? John Dury and the Politics of Irenicism in England, 1628–1643," in *Samuel Hartlib and Universal Reformation: Studies in Intellectual Communication,* edited by Mark Greengrass, Michael Leslie, and Timothy Raylor (Cambridge: Cambridge University Press, 1994). According to notes on Boyle's life made by Sir Peter Pett, Boyle subsidized one of Dury's writings (aimed at reconciling Lutherans and Calvinists); see "Sir Peter Pett's Notes on Boyle," in *Robert Boyle by Himself and His Friends,* edited by Michael Hunter (London: Pickering & Chatto, 1994), pp. 72–73.

isting Socinian works from Latin into English and by writing Socinian works of his own. This systematic presentation of Socinian thought in English prompted a veritable avalanche of rebuttals, one of which Boyle penned but never published, and others that would ultimately influence his thought on "things above reason."

Biddle's translation of Joachim Stegmann's work, first published in 1633, under the English title of *Brevis Disquisitio: Or, a Brief Enquiry Touching a Better way Then is commonly made use of, to refute Papists, and reduce Protestants to certainty and Unity in Religion* was published in 1653. Biddle noted at the very beginning of his introduction, "To the Christian Reader," that

> there will not be wanting some, and they such as profess themselves
> great Zelots in the Protestant cause, who in likelihood will stomach
> at the publication of this little Treatise, because Reason is therein
> (and that not without evident necessity) much cried up.[22]

Citing scriptural support (Romans, chapter 12, verse 1 and I Peter, chapter 2, verse 2) for the rationality of the Gospel,[23] Biddle berated those Protestant ministers who were referring to reason as "Corrupt" and "Carnal," arguing that such ministers enlist reason on their side when it suits them to do so, but denigrate reason when they wish to withhold from rational scrutiny certain "unreasonable" doctrines that they believe. Noting that the Protestant objection to Transubstantiation is that "it is repugnant to *Reason*," he claimed that Transubstantiation has a better claim to a basis in scripture than some other doctrines that were still held by Protestants, and asked why, if "Transubstantiation is to be declaimed because contrary to Reason, . . . shal not all other *Unreasonable* Doctrines upon the same account be exploded." All Christians, Biddle urged, should "set upon a diligent & impartiall triall of all Religious Doctrines by Scripture and Reason, that so we may at length come unto the unity of the Faith."

22. John Biddle, Preface to Joachim Stegmann, *Brevis Disquisitio: Or, a Brief Enquiry Touching a Better way Then is commonly made use of, to refute Papists, and reduce Protestants to certainty and Unity in Religion,* translated by John Biddle (London, 1653), unpaginated.

23. Romans, chapter 12, verse 1, reads, "I beseech you therefore, brethren, by the mercies of God, that ye present your bodies a living sacrifice, holy, acceptable unto God, *which is* your reasonable service." I Peter, chapter 2, verse 2, reads, "As newborn babes, desire the sincere milk of the word, that ye may grow thereby." Biddle argued that the phrase "the sincere milk" should more properly be translated "the sincere rational milk," hence the relevance of this passage.

Biddle's prediction that "great Zelots in the Protestant cause" would object to the Socinian insistence on a rational interpretation of scripture was well-founded, and a look at his translation of Stegmann's *Brevis Disquisitio* reveals why. In it, Stegmann had addressed the questions of what should be the rule of faith, and what should be the judge of that rule. Rejecting the Roman Catholic criteria of papal authority, unwritten tradition, Church Fathers, and councils, Stegmann claimed that for Protestants, scripture alone is the rule of faith. But what, he asked, should be the judge of the rule for Protestants? What would guarantee a correct Protestant interpretation of scripture? Stegmann considered the claim of some Protestants that the Holy Spirit should be the judge, rejecting this criterion on the grounds that it leads to many diverse and contrary opinions, not all of which can possibly be true;[24] who or what, he asked, should be the judge of such apparently conflicting promptings of the Holy Spirit? "The gift of the Holy Spirit doth not take away Reason, but [doth] exalt and enlighten it,"[25] Stegmann claimed. To "know the authority and true meaning of the Scripture, those principles also, which they call Philosophicall, are to be used."[26]

Reason, then, should be the criterion of the proper interpretation of scripture, the standard against which various alternative interpretations are to be measured.[27] Truth should be determined by the forwarding of various learned arguments for the proper interpretation of disputed passages of scripture (such as those concerning the Trinity, the Incarnation, the personal union of the Spirit with the believer, and the state of the soul between the death and resurrection of the body); truth cannot, Stegmann argued, be established by the suppression of dissent. "You [so-called orthodox Protestants]," he complained,

> imagine that you now have the truth, and not the Papists. But others are of a contrary minde. . . . Now by what right do you require of [Papists and other Protestants], that they should think it possible for

---

24. Stegmann, *Brevis Disquisitio*, trans. Biddle, pp. 5–9.
25. Stegmann, *Brevis Disquisitio*, trans. Biddle, p. 11.
26. Stegmann, *Brevis Disquisitio*, trans. Biddle, p. 22.
27. For background information on the problem of the criterion in the sixteenth and seventeenth centuries, see Richard H. Popkin, Introduction to *The Philosophy of the Sixteenth and Seventeenth Centuries*, edited by Richard H. Popkin (New York: Free Press, 1966); see also the essays of Michel Montaigne, Blaise Pascal, and Pierre Bayle included in that anthology. For the relationship between the question of the criterion in theology and in science in the seventeenth century, see van Leeuwen, *The Problem of Certainty*.

them to erre; if you hold it to be impossible for yourselves: But if it
be possible, why suffer you not those that dissent, why do you not
hear them? Why do you prohibit their Books to be read and sold?[28]

Only by submitting alternative interpretations of disputed passages of
scripture to the scrutiny of reason could truth be reached and uniformity
of belief achieved.

Stegmann went on to argue for the truth of certain Socinian beliefs,
the most significant of which for Boyle's *Things above Reason* was the
Socinian interpretation of predestination, a subject I discuss in the next
chapter. The most important thing to note here is that although Stegmann included discussions of doctrinal questions in *Brevis Disquisitio*,
his doing so was clearly peripheral to his primary concern, which was to
establish human reason as the criterion for judging which of alternative
interpretations of scriptural passages should be accepted as correct.

The systematic application of Stegmann's criterion is apparent in the
Racovian Catechism, translated by Biddle in 1652.[29] For each of the
doctrinal questions considered, Biddle cited both scriptural passages that
could be used to support the Socinian position and scriptural passages
that could be used to deny the Socinian position. After submitting the
various passages to reason's scrutiny, Biddle argued that individuals
who attempted to refute Socinianism by relying on passages apparently
contradictory to Socinian doctrines were appealing to passages "ill understood."[30] Christ's divinity, for example, was denied first on scriptural
grounds, and then on the grounds that any other passages in scripture
that appear to support Christ's divinity should not be so interpreted
because such an interpretation is repugnant to reason – that is, two

28. Stegmann, *Brevis Disquisitio,* trans. Biddle, p. 45.
29. *Catechism of the Church of those people who, in the Kingdom of Poland
    and the Grand Duchy of Lithuania and in other Domains belonging to the
    Crown, affirm and confess that no other than the Father of our Lord Jesus
    Christ is the only God of Israel and that the Man Jesus the Nazarene, who
    was born of the Virgin, and no other besides him, is the only begotten Son
    of God* (in Polish, Raków, 1605) was published by Biddle with the title of
    *The Racovian Catechisme: Wherein you have the substance of the Confes-
    sion of those Churches, which in the Kingdom of Poland, and Great
    Dukedome of Lithuania, and other Provinces appertaining to that king-
    dom, do affirm, that no other save the Father of our Lord Jesus Christ, is
    that one God of Israel, and that the man Jesus of Nazareth, who was born
    of the Virgin, and no other besides, or before him, is the onely begotten
    Sonne of God* (Amsterledam [London], 1652). All subsequent references to
    the *Racovian Catechism* are to Biddle's edition.
30. *Racovian Catechism,* pp. 18–19.

persons imbued with opposing qualities (to be both mortal and immortal, to have a beginning and to be eternal, to be both mutable and immutable) cannot constitute one substance.[31] In short, passages traditionally invoked in support of orthodox Trinitarianism should be reexamined; in the case of John, chapter 1, verse 14, for example, Biddle argued that the original Greek should be translated "The Word was flesh," rather than "The Word was made flesh."[32]

I turn now to another Socinian catechism, one penned by Biddle himself and published in 1654. This catechism, *A Twofold Catechism*, differs from the *Racovian Catechism*; in it Biddle did not address himself to both Socinian and anti-Socinian points of view, but instead patterned the work on more traditional catechisms by simply posing questions and then answering them.[33] The answers were, of course, an exposition of Socinian doctrine and were supported by appeals to scripture; Biddle offered no argument and no commentary. In denying predestination and original sin and in affirming man's free will, for example, Biddle asked, concerning the freely chosen actions of human beings

> which are neither past, nor present, but may afterwards either be or
> not be, what are the chief passages of the Scripture, from whence it
> is wont to be gathered, that God knoweth not such actions, till they
> come to pass? yea, that there are such actions?[34]

This question was followed by six pages of scriptural citations from both the Old and the New Testaments offered as proof that God does *not* have foreknowledge of future contingents that result from man's free will. For example, God brought the beasts to Adam to see what Adam would call them (Genesis, chapter 2, verse 19), and Christians are ex-

---

31. *Racovian Catechism*, p. 28. The scriptural passages adduced in support of Socinianism were I Timothy chapter 2, verse 5 ("For *there is* one God, and one mediator between God and men, the man Christ Jesus") and I Corinthians, chapter 15, verse 21 ("For since by man *came* death, by man *came* also the resurrection of the dead").

32. *Racovian Catechism*, p. 53.

33. This fact is reflected in the full title of the work: *A Twofold Catechism: The One simply called A Scripture-Catechism; the Other, A brief Scripture-Catechism for Children; wherein the Chiefest points of the Christian Religion, being Question-wise proposed, resolve themselves by pertinent Answers taken word for word out of the Scripture, without either Consequences or Comments. Composed for their sakes that would fain be Meer Christians, and not of this or that Sect, inasmuch as all the Sects of Christians, by what names soever distinguished, have either more or less departed from the simplicity and truth of the Scripture* (London, 1654).

34. Biddle, *Two-Fold Catechism*, p. 14.

horted to let their requests be known to God (Philippians, chapter 4, verse 6).

Delving further into Socinian doctrines would be interesting, but it is not relevant here. The primary relevance of the writings of the Socinians to Boyle's views was their insistence that human reason must be the criterion against which proposed scriptural interpretations be measured, for this is the claim that Boyle would deny in *Things above Reason*. Approximately three decades prior to the publication of *Things above Reason*, however, he wrote "Essay of the holy Scriptures," parts of which were later published as *Style of the Scriptures* (1661), other parts of which found their way into *Some Physico-theological Considerations about the Possibility of the Resurrection* (1675), and some parts of which he never published. His initial reaction to the ideas of the Socinians is contained in the never-published sections.

Boyle's Response to Socinianism (c. 1652)

Boyle may have been introduced to Socinian writings by John Dury,[35] or by James Ussher, Archbishop of Armagh[36] and a mentor of Boyle's, who had met with the English Socinian John Biddle in an attempt to convince Biddle of the error of his ways, or by Benjamin Worsley, another mentor of Boyle's known to have collected Socinian works.[37] In any event, he revealed his familiarity with Socinian writings in the "Essay of the holy Scriptures," most likely written around 1652.[38]

35. The latest date by which Dury was known to Boyle was 1645; see R.E.W. Maddison, "Studies in the Life of Robert Boyle, F.R.S.: Part VI, The Stalbridge Period, 1645–1655, and the Invisible College," *Notes and Records of the Royal Society of London*, 18 (1963), p. 105.

36. For background on Ussher, see Trevor-Roper, *Catholics, Anglicans and Puritans*, pp. 120–165.

37. For the possibility of Ussher's or Worsley's influence, see Michael Hunter, "How Boyle Became a Scientist," *History of Science* 33 (1995), pp. 72, 80–83.

38. The manuscript is in an unidentified hand, and the exact date of composition is not known, although at least parts postdate the publication of J.C. Hottinger's 1651 *Historia Orientalis*, mentioned on fols. 47, 65. References to other "recent" works are less conclusive because the exact works or editions to which Boyle referred are not known. Mention of a work in which Socinians assent to the resurrection of the body, for example, could be a reference to chapter eight of John Biddle's translation of Joachim Stegmann's *Brevis Disquisitio* (1653), although the Latin edition of this

"Essay of the holy Scriptures" is a ninety-four page manuscript that is an early version of Boyle's *Style of the Scriptures*. In it, Boyle addressed many of the same issues he would later discuss in *Style;* indeed, some passages from the manuscript are to be found *verbatim* in *Style*.[39] The manuscript, however, contains a direct attack against the Socinians that is not included in *Style,* and it also contains a rudimentary version of Boyle's views on the proper relationship of reason to revelation, views he would develop more fully in his later writings, and especially in his *Discourse of Things above Reason.*

The "Essay" began with an argument (aimed at Roman Catholics) that not only should each Christian have access to scripture in the vernacular, but also that each Christian has a positive duty to study scripture. Then Boyle went on to note that the proper interpretation of scripture poses a problem because individuals tend to interpret scripture in accordance with their own preconceived notions.[40] It is not unknown, Boyle continued, how

> clearly against the Light of Reason the Article of the Trinity seems
> to those that deny it; that the Resurrection of the Dead seems to

---

work was widely available in England by 1641. Boyle also quotes a passage from Faustus Socinus's second letter to Andreas Dudithius; this passage can be found in *Bibliotheca Fratrum Polonorum: Socini Opera* (Irenopoli [Amsterdam], 1656), vol. 1, p. 498. This does not decisively date the manuscript, however, because the letter had been previously published in *Fausti Socini Senensis ad Andream Dudithium Epistolæ ex Italico in Latinum conversæ à M.R.H.* (Racoviæ, 1635), pp. 16–35; the passage Boyle quoted is on p. 19. Boyle's statement in *Style* that he had composed a longer version of the work some nine or ten years earlier would seem to indicate that the "Essay" was composed around 1651 or 1652. For an excellent discussion of this essay, see Hunter, "How Boyle Became a Scientist," esp. pp. 71–77; additional dating clues may be found in Hunter's notes 51 and 108.

39. The relationship between the manuscript and *Style* is interesting; passages do not always turn up in *Style* in the order in which they occur in the manuscript, and much material has been added in *Style* that does not occur in the manuscript at all; compare, for example, fol. 6 with *Style, Works,* vol. 2, pp. 267–268, fol. 8 with p. 267, fol. 9 with p. 268, fols. 13–14 with p. 265; and fols. 14–19 with pp. 281–283. Some material in the manuscript that does not occur in *Style* occurs elsewhere in Boyle's published works. The passage from Socinus's letter to Dudithius (see note 38), for example, may be found in *Reason and Religion, Works,* vol. 4, p. 172, and fols. 57–70 are closely related to the material in *Possibility of the Resurrection, Works,* vol. 4, pp. 191–202.

40. Boyle Papers, vol. 7, fols. 15–16.

many an Absolute Impossibility, . . . [and] that the Life of the Soule seperate from the Body betwixt the Days of Death & Judgment seemes to divers modern Socinians & others soe little to be conceiv'd that 'tis not at all to be beleev'd.[41]

His target here was the Socinians, and he revealed a familiarity with their writings when he added that the "Socinians, who all of them deny the Trinity, & divers of them the Sensibility of the Seperate Soule, are infinitely subtle Men, & incomparable Masters of Reason," admitting that he had, on occasion, learned from them. In fact, he had "sometimes in the writings of Socinus, & some of his Adherents, met Arguments so dexterously manag'd as to make [him] wish the Truth had numbred those skillfull Fencers amongst hir Champions."[42] Nevertheless, even when the Socinians, the more orthodox Protestants, and the Roman Catholics, who were agreed on the basic truth of the "Grand Fundamentalls of Christianity" joined forces to prove the truth of Christianity against the "Atheists, Jews & other Infidells," Boyle did not think the arguments of "Socinus, Crellius, Volkelius, & the Racovian Catechists" were more compelling than those who professed the truth of Christianity in more orthodox terms than did the Socinians.[43]

Although in Boyle's view it was the duty of each Christian to study scripture – and in Boyle's opinion learned exegesis of scripture could aid in understanding[44] – he stressed that human reason is but an aid to proper understanding and not the criterion against which revelations should be judged. "Those that will rightly judge of Truth must be Inquisitive Examiners but not Refractory," he argued, and noted that this is particularly true where revelations concerning the soul and the Trinity are concerned, because God's ways are so incomprehensible that human beings have problems understanding these difficult revelations.[45] Whatever God reveals is true, he argued, and Christians simply must accept those revelations as true. "I could wish," he continued, that "men had bin lesse Positive & Determinate in their Discourses about the Trinity; in which a Reverence is safer than Nicety."[46]

41. Boyle Papers, vol. 7, fol. 23.
42. Boyle Papers, vol. 7, fol. 45.
43. Boyle Papers, vol. 7, fol. 46.
44. For a discussion of Boyle's interest during this period of his life in the philological learning of the day, an interest that led him to the study of Greek, Hebrew, and other Biblical languages, see Hunter, "How Boyle Became a Scientist," pp. 71–73.
45. Boyle Papers, vol. 7, fol. 47.
46. Boyle Papers, vol. 7, fol. 48.

Where such difficult doctrines are concerned, Boyle warned that Christians should not allow the improbability of the thing – or even its apparent impossibility – to impede belief, especially when God's omnipotence is taken into consideration.[47] After all, he continued, some inability to understand fully is essential to faith,[48] and doctrines such as those concerning the Trinity, the resurrection of the dead, and the existence of the soul between the death and resurrection of the body, even though difficult to understand, must be true because they are attested to by miracles.[49] The role of reason in such controversies, Boyle argued, is "not to demonstrate the Truth of each particular Article of Christianity," but instead to judge the evidence (such as miracles) that the scriptures are the revealed word of God.[50]

It is important to note several things about this early version of *Style of the Scriptures*. First, Boyle clearly believed that the Socinians were doctrinally in error where certain of their beliefs (such as the insensibility of the soul between the death and resurrection of the body and their anti-Trinitarianism) were concerned. Second, even in this relatively early writing, Boyle emphasized the limits of human understanding, and he did so in response to the Socinian claim that reason is to be the criterion against which scriptural revelations should be interpreted. And third, although some of the passages I have quoted are among the most polemical ones to be found in Boyle's writings, published or unpublished, Boyle never relinquished his gentlemanly style. Although his "love of truth" led him to oppose the views of the Socinians in this draft version of *Style*, his "love of peace" led him to suppress passages in the published version that could be considered polemical and that might contribute to doctrinal divisiveness.[51]

47. Boyle Papers, vol. 7, fol. 17. In the manuscript, this sentence is marked "to be deleted," although later, in *Things above Reason*, Boyle would argue just this point (as I show in chapter 4).

48. Boyle Papers, vol. 7, fol. 18.

49. Boyle Papers, vol. 7, fols. 25–26.

50. Boyle Papers, vol. 7, fols. 26–27. In the remainder of this manuscript, Boyle devoted fourteen folios to arguing that the resurrection of the body is not impossible when God's omnipotence is taken into consideration; this section closely resembles the argument Boyle later published as "Physico-Theological Considerations concerning the Resurrection of the Body," although to date I have been unable to find any passages in the latter that appear *verbatim* in the manuscript. Then, in the final folios, Boyle addressed questions related to the fulfillment in the New Testament of Old Testament prophecies.

51. I discussed possible reasons for the suppression of these passages in the Introduction (in the section "Boyle's 'Love of Peace' and 'Love of Truth' ").

Boyle would later develop more fully and systematize his views on reason's limits, especially in the *Discourse of Things above Reason* and its accompanying *Advices in judging of Things said to transcend Reason* (1681), as well as in *Reflections upon a Theological Distinction,* which was appended to *The Christian Virtuoso* (1690). But between this early "Essay of the holy Scriptures," in which his expressions of reason's limits were sketchy and not supported by any detailed argument, and his full exposition of reason's limits in *Things above Reason,* there were a variety of responses to Socinianism penned by his contemporaries. In them can be found all of the concepts that formed various aspects of Boyle's mature thought. It is to those I now turn.

Other Responses to Socinianism

Those who wished to rebut the Socinian claim that reason should be the criterion of conflicting interpretations of scripture could do so in two different ways. Either they could deny that reason should in fact be the criterion, or they could agree with the Socinians that indeed reason *should* be the criterion (but argue that the Socinians had reasoned poorly and missed the mark in their own rational interpretations). John Owen (in his *Vindiciæ Evangelicæ* of 1655) and Joseph Glanvill took the latter approach, Owen implicitly and Glanvill explicitly. Richard Baxter, Robert Ferguson, and (again) John Owen (in his much later *Nature of Apostasie,* 1676) took the former approach.

John Owen's Vindiciæ Evangelicæ

John Owen, a high Calvinist, was a religious advisor to Oliver Cromwell during the Interregnum, and the leader of the Independents after the Restoration.[52] Apparently he had been reading Socinian works and had been concerned about the threat of Socinianism for some time prior to Biddle's making Socinian views widely available in English in 1653 and

---

Although Boyle's suppression of the passages might be considered irenical, the "Essay" itself is not, as James R. Jacob would have it, "expressly irenical" (*Robert Boyle and the English Revolution: A Study in Social and Intellectual Change* [New York: Burt Franklin & Co., 1977], p. 26).

52. For background information on Owen, see Peter Toon, *God's Statesman: The Life and Works of John Owen, Pastor, Educator, Theologian* (Exeter: Paternoster Press, 1971). See also Alan C. Clifford, *Atonement and Justification: English Evangelical Theology 1640–1790. An Evaluation* (Oxford: Clarendon Press, 1990), pp. 3–16.

1654.[53] In 1654, the Council of State asked him to write (in English) a
rebuttal of Socinianism, and in the full title of *Vindiciæ Evangelicæ or
The Mystery of the Gospell Vindicated, and Socinianisme Examined*,
published in 1655, Owen made it clear that he was writing in response
to Biddle's *Scripture Catechisme*, and to "the Catechisme of Valentinus
Smalcius, commonly called the *Racovian Catechisme*."[54] In the separately paginated seventy-page preface, Owen showed that he knew well
the history of various forms of anti-Trinitarianism, including the form
he was specifically concerned to rebut, the recent claim of the Socinians
that Jesus Christ was merely human. In the body of the work, Owen
revealed his familiarity with the Latin writings of the Socinians; he cited,
for example, Volkelius's *De vera religione* (p. 47), Schlictingius's letter to
Meisner (p. 47),[55] Crellius's *De deo et eius attributis* (p. 22), and Smalcius's *De divinitate Jesu Christi* (p. 47).

Although Owen noted and rejected the Socinians' "high *pretences to
Reason*,"[56] it was obvious that he believed that the Socinians should be
attacked on their own grounds. "Let not any *attempt dealing* with these
men," Owen urged,

> that is not in some *good measure* furnished with those kinds of
> *literature*, and those *common Arts*, wherein they excell: as first, the
> *knowledge* of the *Tongues*, wherein the *scripture* is *written*, . . . [as
> well as] *Logick* and *Rhetorick*.[57]

---

53. See, for example, Owen's *Diatriba de justitia divina: Seu justitiæ vindica-
tricis vindiciæ* (Oxford, 1653), in which he cited most of the major Socinian works. Owen's own library contained at least thirty-three Socinian
works (McLachlan, *Socinianism in Seventeenth-Century England*, pp. 128–
129).

54. The full title of Owen's work is *Vindiciæ Evangelicæ; or, The Mystery of
the gospell vindicated, and Socinianisme Examined, in the Consideration,
and Confutation of a Catechisme, called A Scripture Catechisme, Written
by J. Biddle M. A., and the Catechisme of Valentinus Smalcius, commonly
called the Racovian Catechisme, with the Vindication of the Testimonies of
Scripture, concerning the Deity and Satisfaction of Jesus Christ, from the
Perverse Expositions, and Interpretations of them, by Hugo Grotius in his
Annotations on the Bible; also an Appendix, in Vindication of some things
formerly written about the Death of Christ, & the fruits thereof, from the
Animadversions of Mr. R[ichard] B[axter]* (London, 1655).

55. This is either Johannes Schlichtingius's *Quæstio num ad regnum Dei* ([Ra-
ków], 1635), his *Quæstiones . . . contra Balthsarem Meisnerum* ([Raków]
1636), or *Ionæ Schlichtingii à Bukowiec . . . Adversùs Balthasarem Meisnerum* ([Raków], 1637).

56. Owen, *Vindiciæ Evangelicæ*, preface, p. 63.

57. Owen, *Vindiciæ Evangelicæ*, preface, pp. 65–66.

That Owen considered himself well-furnished with the skills needed to rebut the claims of the Socinians is illustrated throughout the work; for example, Owen reprinted Biddle's four-and-a-half-page preface to the *Scripture Catechism* at the beginning of the main text of *Vindiciæ Evangelicæ*, and then devoted almost forty pages to a point-by-point rebuttal of that preface.

An examination of only a few of Owen's arguments will illustrate the wide range of weapons with which he opposed the Socinian threat. For example, in asserting what he considered to be the "glorious truth," the eternal deity of Jesus Christ, he rejected the Socinian interpretation of the first chapter of John, arguing that there was confusion concerning the nature of the Trinity even before John was written, and that in the first chapter of John, God had explicitly addressed himself to this confusion and revealed the true nature of the Trinity, especially in verse 1, which reads, "In the beginning was the Word, and the Word was with God, and the Word was God," and in verse 14, which reads, "And the Word became flesh and dwelt among us." Instead of twisting the interpretation of these passages as the Socinians were doing, Owen complained, Christians should accept God's word on the matter as final.[58]

Next, Owen argued that the devil was responsible for the recent revival of anti-Trinitarian arguments, claiming that in the early days of Christianity, Satan had tried to destroy the Church by spreading anti-Trinitarian heresies. Having failed to subvert the early Church with such heresies, Satan had busied himself with other things. Now, in the wake of the Reformation, Satan had a new target, and was trying to subvert the Reformed Church by once again spreading anti-Trinitarian heresies.[59]

Finally, Owen argued that the Socinians were in no position to complain that the orthodox interpretation of the Trinity was contrary to reason because some Socinians held contradictory views regarding Christ's status as a mere man themselves. Faustus Socinus himself, Owen noted, had claimed that Jesus Christ was merely human *and* that Christ should be invoked in prayer and worshipped; surely, he pointed out, this was a contradictory position to hold.[60]

Throughout *Vindiciæ Evangelicæ* Owen implicitly assumed that he could apply his knowledge of ancient tongues, logic, rhetoric, and ratio-

58. Owen, *Vindiciæ Evangelicæ,* preface, pp. 5–6. For the Socinian interpretation of John, chapter 1, verse 14, see p. 54.
59. Owen, *Vindiciæ Evangelicæ,* preface, pp. 8–9.
60. Owen, *Vindiciæ Evangelicæ,* preface, pp. 30–31.

nal argumentation to doctrinal questions more skillfully than the Socinians could, and that in so doing he could prove their scriptural interpretations to be erroneous and his own interpretations to be correct. Although in his later anti-Socinian work (to be examined later) Owen argued that human reason is not competent to comprehend fully the content of revelation, nowhere in *Vindiciæ* did he place any sustained emphasis on reason's limits.

Richard Baxter

Richard Baxter was a moderate Presbyterian whose doctrinal beliefs were in many ways closer to those of the latitudinarians than they were to those of high Calvinists such as Owen, although Baxter preferred to be known as a "reconciler" unassociated with any particular party.[61] In the same year that Owen published *Vindiciæ*, Baxter published *The Unreasonableness of Infidelity: Manifested in four discourses*; the fourth part, *The Arrogancy of Reason against Divine Revelations, Repressed; or, Proud Ignorance the cause of Infidelity, and of Mens Quarrelling with the Word of God*, was aimed (in part) at the Socinians. In *Arrogancy*, Baxter, unlike Owen, emphasized reason's limits and argued that human reason should *not* be the criterion against which different interpretations of scripture should be evaluated, although he revealed a certain tension in his thought concerning reason's role when he (like Owen) used reason to rebut some of the doctrines he considered to be erroneous.

In *Arrogancy*, Baxter appealed to John, chapter 3, verse 9 (in which Nicodemus asks Christ "How can these things be?") to argue that

> the corrupt nature of man is more prone to question the truth of Gods Word, then to see and confess their own ignorance and incapacity; and ready to doubt, whether the things that Christ revealeth are true, when they themselves do not know the nature, cause, and reason of them.[62]

---

61. For background information on Baxter, see Irvonwy Morgan, *The Non-conformity of Richard Baxter* (London: Epworth Press, 1946); see also Clifford, *Atonement and Justification*, pp. 17–31; Isabel Rivers, *Reason, Grace, and Sentiment: A Study in the Language of Religion and Ethics in England, 1660–1780*, vol. 1, *Whichcote to Wesley* (Cambridge: Cambridge University Press, 1991), chapter 3.

62. Richard Baxter, *The Unreasonableness of Infidelity: Manifested in four discourses*; the fourth part, *The Arrogancy of Reason against Divine Revelations, Repressed; or, Proud Ignorance the cause of Infidelity, and of Mens Quarrelling with the Word of God* (London, 1655), p. 13. *The Arrogancy*

In some cases, such as the nature of the immortal soul or the nature of the believer's communion with God, Baxter complained, some individuals refuse to assent to a revealed truth because they do not understand the nature (or "quiddity") of the thing revealed.[63] In other cases, individuals refuse to assent to a revealed truth because they do not understand how the thing revealed can be caused; Baxter's examples of such revealed truths included predestination and the creation of the world. In still other cases, individuals refuse to assent to a revealed truth (such as Christ's Satisfaction and its universal extent) because they do not understand God's ends and reasons.[64] And sometimes individuals reject a revealed truth simply because they cannot answer all of the objections that can be raised against that truth. But, Baxter argued, human knowledge is imperfect, especially where the mysteries of theology are concerned, "and doubtless where men are so defective in knowledge, there must still be difficulties in their way, and many knots which they cannot untie."[65]

In the course of his discussion, Baxter explicitly addressed the question of apparent contradictions in scripture, noting that some individuals reject Christianity because they believe that the Christian revelation contains contradictions.[66] In responding to this claim, Baxter denied that scripture contains contradictions. In the process, however, he emphasized the limits of human understanding in such a way as to make his argument as a whole ambiguous. First, he argued

> that no man can know Gods truths perfectly, till he see them all as in one Scheam or Body, with one view, as it were, and so sees the Location of each Truth, and the respect that it hath to all the rest; not onely to see that there is no contradiction, but how every Truth doth fortify the rest. All this therefore is exceeding desirable, but it is not every mans lot to attain it, nor any mans in this world perfectly, or near to a perfection.[67]

At this stage of his argument, it might seem as if he were about to argue that scripture indeed contains contradictions from the point of view of imperfect human understanding, and indeed he went on to argue that even after the "best men" have studied the languages in which the scriptures were originally written and their historical context, those men

---

of Reason was also published as a separate work the same year; the pagination of the two editions is the same.

63. Baxter, *Arrogancy*, p. 16.
64. Baxter, *Arrogancy*, pp. 18–20.
65. Baxter, *Arrogancy*, p. 27.
66. Baxter, *Arrogancy*, pp. 21–23.
67. Baxter, *Arrogancy*, p. 23.

"will here know [truth] but in part."[68] Further, at the end of *Arrogancy* he went into considerable detail about alleged contradictions in scripture when he dealt with the objection that to believe something that is unreasonable is to make us "mad, and not Christians." Here he began again by stressing reason's limits. Those who make such an objection, Baxter noted, argue that "it is certain that God never spoke contradictions. Therefore if I finde contradictions in the Scriptures, may I not rationally argue that they are not the Word of God?" To this he responded that the minor premise (that there are contradictions in scripture) could never be proved because of our "shallow understanding."[69] However, he went on, somewhat illogically, to show that human reason *is* competent to prove that there are *no* contradictions in scripture, for "God attesteth no contradictions: but God attesteth the holy Scripture. Therefore the holy Scriptures have no contradictions."[70]

Baxter could have connected his arguments concerning reason's limits and reason's competence in such a way as to avoid any ambiguity had he argued explicitly that God, being incapable of contradiction, would attest to no contradictions in scripture, but that to circumscribed human understanding there might *appear* to be contradictions in scripture (a move that Boyle would make in *Things above Reason*). Baxter did not make this move; rather, he went on to consider the specific allegation that Christians assert a contradiction when they assert that there are three persons in the Godhead but only one God. Instead of arguing that the doctrine of the Trinity is one of those "knots" that human reason cannot "untie," he attempted to "untie the knot" by arguing that the orthodox interpretation of the Trinity does not involve a contradiction because the three persons in the Trinity should be conceived as analogous to the vegetative, the sensitive, and the rational soul of human beings – each is distinct, yet there is only one soul.[71]

This tension in Baxter's thought is reflected in his subsequent writings on contradictions in Christian doctrines. In *Reasons of the Christian Religion*, he denied that scripture contains any contradictions (at least in things necessary for salvation) because in such matters the Holy Spirit infallibly guides the human authors of the scriptures.[72] In his *Judgment of Non-Conformists, of the Interest of Reason, in Matters of Religion,*

68. Baxter, *Arrogancy*, p. 35.
69. Baxter, *Arrogancy*, p. 60.
70. Baxter, *Arrogancy*, pp. 60–61.
71. Baxter, *Arrogancy*, p. 75.
72. Richard Baxter, *The Reasons of the Christian Religion* (London, 1667), pp. 412–415.

however, Baxter reverted to the original point of *Arrogancy* – that is, that human understanding is incapable of perceiving the consistency ("harmony") among all of God's truths. Here Baxter distinguished between human reason first as it was in the state of innocence, then as debased (as it was by Adam's fall), and finally as "sanctified" or "recovered" as it is in the believer. In the state of innocence, Baxter argued, human reason could understand both "natural, and supernatural manifestations of Gods governing Will." Debased reason, however, judges "carnally, falsely, and malignantly" of spiritual things. When reason is sanctified by belief in Christianity, the Holy Spirit helps to guide human reason toward a correct understanding of the scriptures.[73] Even when human reason is sanctified, however, it is not restored to the perfect state it enjoyed in the state of innocence; once debased, human reason will always be to some extent contrary to religion. Only the blessed in glory, Baxter argued, will perceive the harmony among all of God's truths. When individuals encounter passages in scripture they cannot reconcile with other passages or with other known truths, Christians should accuse "their [own] dark, and erring understandings" rather than denying the truths revealed in those passages.[74]

Joseph Glanvill

As we have seen, John Owen had implicitly adopted the Socinians' criterion for the interpretation of scripture in *Vindiciæ*, when he had attempted to show that human reason, aided by such tools as logic, rhetoric, and a knowledge of the languages in which scripture was originally written, could be used to confute Socinian doctrines. In 1670, Joseph Glanvill, Anglican divine and latitudinarian, published ΛΟΓΟΥ ΘΡΗΣΚΕΙΑ; *or, A Seasonable Recommendation, and Defence of Reason, in the Affairs of Religion; Against Infidelity, Scepticism, and Fanaticisms of all sorts,* in which he argued explicitly that human reason must be accepted as the criterion against which various interpretations of scripture should be judged in order to refute the erroneous doctrines of the Socinians.[75]

In *Seasonal Recommendation,* Glanvill noted that "'tis from the *Pul-*

---

73. Richard Baxter, *The Judgment of Non-Conformists, of the Interest of Reason, in Matters of Religion* (London, 1676), pp. 8–10.
74. Baxter, *Judgment of Non-Conformists,* pp. 15, 17.
75. For background information on Glanvill, see Jackson I. Cope, *Joseph Glanvill: Anglican Apologist* (St. Louis: Washington University Press, 1956).

pit, Religion hath received *those wounds* through the *sides* of *Reason,"*
and then went on to complain that such clergymen "set up a *loud* cry
against *Reason* ... under the misapplied names of *Vain Philosophy,
Carnal Reasoning,* and the *Wisdom of this World."*[76] This denigration
of human reason, he claimed, had contributed to those

> *sickly conceits,* and *Enthusiastick dreames,* and *unsound Doctrines,*
> that have poysoned our Aire, and infatuated the minds of men, and
> exposed *Religion* to the *scorn* of *Infidels,* and *divided* the *Church,*
> and disturbed the *peace* of mankind, and involved the *Nation* in so
> much *bloud,* and in so many *Ruines.*[77]

Making it clear that he was aware that some of his contemporaries
charged that "'tis Socinianism to plead for Reason in the affairs of
Faith, and Religion," Glanvill argued that it was absurd to suppose that
Socinians were

> the only *rational* men; when as divers of their Doctrines, such as,
> The *Sleep,* and *natural mortality* of the *soul,* and utter *extinction,*
> and *annihilation of the wicked after the day of judgment,* are very
> *obnoxious* to *Philosophy,* and *Reason.* And the *Socinians* can never
> be confuted in their *other* opinions without using *Reason* to main-
> tain the *sense,* and *interpretations* of those *Scriptures* that are
> alledged against them.[78]

In developing his argument, Glanvill divided Christianity into two
parts. The first part consisted of those things knowable by reason alone:
God's existence and the divine authority of scripture, the latter being
attested to by miracles, the excellency of the Christian doctrine, the
influence of Christianity on the souls of human beings, and God's providence in preserving the Christian religion. In the second part, he placed
the formal articles of faith expressly declared in scripture. These formal
articles were in turn divided into two kinds, "mixt" and "pure." The
mixed articles were those Glanvill claimed to be both discoverable by
reason and revealed in scripture; he included among these a knowledge

---

76. Joseph Glanvill, ΛΟΓΟΥ ΘΡΗΣΚΕΙΑ; *or, A Seasonable Recommendation,
and Defence of Reason, in the Affairs of Religion; Against Infidelity,
Scepticism, and Fanaticisms of all sorts* (London, 1670), p. 2. There are
two 1670 editions of Glanvill's *Seasonable Recommendation.* One edition
is continuously paginated with his *Philosophia Pia* (although it is separately titled); in the other edition *Seasonable Recommendation* appears
alone. I cite the latter. The essay was republished in a slightly revised
version as "The Agreement of Reason and Religion," in Glanvill's *Essays
on Several Important Subjects in Philosophy and Religion* (London, 1676).
77. Glanvill, *Seasonable Recommendation,* p. 1.
78. Glanvill, *Seasonable Recommendation,* pp. 22–23.

of God's attributes, of the existence of moral good and evil, and of the immortality of the human soul. The pure articles, on the other hand, could be known only by divine testimony as revealed in scripture. In this group he included the doctrines of the Miraculous Conception, the Incarnation, and the Trinity, and argued that even though human reason is incapable of understanding *how* these mysteries could be true, they must be believed because reason should compel assent to what God says is true.[79]

It was in considering the claim that some parts of the Christian revelation are "above reason" that Glanvill argued that "no more is meant [by the claim], than that *Reason* cannot *conceive how those things are*," noting that "in *that* sense many of the affairs of *nature* are *above it* too," (such as how the parts of matter cohere and how the human soul is united to the human body).[80] His statement that it is reasonable to believe the mysteries of Christianity simply because God has revealed them, even though we cannot understand *how* they are true, is reminiscent of Baxter's point concerning reason's *limits* in *Arrogancy*, in which he had appealed to Nicodemus's question, "How can these things be?" That Glanvill's emphasis was on reason's *competence* rather than its limits, however, is evident in his discussion of "Inferences that may be raised from the whole [argument]," in which he concluded that "no Principle of Reason contradicts any Articles of Faith." Both the principles of reason and the articles of faith, he claimed, are gifts of God, and as such, cannot clash. They may sometimes *seem* contradictory, but in those cases, "either something is taken for *Faith*, that is but a *Phancy*; or something for *Reason*, that is but *Sophistry*; or the *supposed contradiction* is an *error*, and *mistake*."[81]

"When any thing is pretended from Reason, against any Article of Faith," he continued, "we ought not to cut the knot, by denying reason; but indeavour to unty it by answering the Argument, and 'tis certain it may be fairly answered," because all heretics either argue from false principles or reach fallacious conclusions from true ones, and therefore their mistakes could be discovered. "When any thing is offered us for an Article of Faith that seems to contradict Reason," Glanvill added, "we ought to see [if] there be good cause to believe that this is divinely revealed, and in the sense propounded." If so, then the contradiction is but apparent and the faulty reasoning behind it can be exposed. Should the contradiction turn out to be real, however, then the pro-

---

79. Glanvill, *Seasonable Recommendation*, pp. 8–13.
80. Glanvill, *Seasonable Recommendation*, p. 14.
81. Glanvill, *Seasonable Recommendation*, p. 25.

posed article cannot be considered a genuine article of faith and the
revelation is being misinterpreted, for God cannot be the author of contradictions.[82] Nowhere did Glanvill suggest that the harmonious (or
noncontradictory) nature of divine revelations might be comprehended
only by God, or (as Baxter had put it) that human beings might be
capable of understanding the consistency of all of God's truths only in
heaven.

Despite his insistence that human reason could "untie" doctrinal
knots, Glanvill did not attempt to show that the doctrine of the Trinity
is consistent with human reason (as Baxter had done); in fact, he did not
discuss in detail any disputed doctrine. Indeed, although he had included
the doctrine of the Trinity (as well as the doctrines of the Miraculous
Conception and Incarnation) among the articles of Christianity in his
discussion of "mixt" and "pure" revelations, his actual definition of
Christianity was rather minimalist. He explicitly defined Christianity as
consisting of a belief in Baptism, the Lord's Supper, God's commandments, and the Apostles' Creed (in addition to those things knowable by
reason alone, such as knowledge of God's existence and the existence of
moral good and evil).[83] Nowhere did he refer to the Thirty-Nine Articles
of the Church of England, in which doctrines related to original sin,
predestination, justification, and many other matters are spelled out. (I
should note that he did touch rather lightly on some aspects of predestination, such as that if sinners repent God will forgive them, in his
discussion of the duties of religion; however, he classified these beliefs as
"incouraging" and helpful, not "fundamental."[84]) Further, the only
creed he mentioned was the Apostles' Creed, despite the fact that the
Eighth Article of the Church of England required assent to the Nicene
and Athanasian Creeds as well, both of which, unlike the Apostles'
Creed, assert the doctrine of the consubstantiality of the three persons
of the Trinity, as well as the co-eternality of the Son with the Father. (It
is interesting to note that in his listing of the erroneous doctrines of the
Socinians quoted earlier, Glanvill did not mention specifically their anti-
Trinitarianism.) Regardless of the significance – or lack thereof – of
Glanvill's reluctance to discuss doctrinal details, one thing is crystal
clear: He was arguing that human reason is competent to interpret
revelation and that consonance with human reason is the criterion
against which differing interpretations of scripture should be judged.

82. Glanvill, *Seasonable Recommendation*, pp. 25–26.
83. Glanvill, *Seasonable Recommendation*, pp. 5–7.
84. Glanvill, *Seasonable Recommendation*, p. 5.

Robert Ferguson

In 1675, Robert Ferguson, a nonconformist, locked horns with Glanvill in *The Interest of Reason in Religion . . . and the Nature of the Union Betwixt Christ & Believers*. Ferguson, who was at that time John Owen's assistant at Owen's Independent (Congregational) church in London, had two main reasons for writing *Interest*. The first was to rebut Glanvill's claims in *Seasonal Recommendation* concerning reason's competence to judge of revealed doctrines. The second was to deny the claim of William Sherlock, an Anglican clergyman, that the union of Christ with believers was mediated by the Church (a claim Sherlock had made in *A Discourse Concerning the knowledge of Jesus Christ, and Our Union and Communion with Him*, published in 1674); in *Interest*, Ferguson argued that Christ's union with individual believers is immediate. I discuss Ferguson's argument against Glanvill first.

Ferguson began by identifying himself as the target of Glanvill's attack in *Seasonable Recommendation*, quoting the passage in which Glanvill had berated those who "set up a *loud* cry against *Reason* . . . under the misapplied names of *Vain Philosophy, Carnal Reasoning*, and the *Wisdom of this World*."[85] He responded to this charge by devoting the first 274 pages of his work (which runs to 665 pages altogether) to an exposition of his view of the proper relationship of reason to the content of revelation. As had Glanvill, Ferguson denied that any revealed truths might be contrary to reason; unlike Glanvill, however, Ferguson distinguished between "universal" (or "abstract") reason and reason as depraved by Adam's fall and exercised by humans ("concrete" reason). It is depraved reason, Ferguson explained, that can legitimately be charged with being "unfriendly" to religion.[86]

In his discussion of Glanvill's category of "pure" revelations (those truths that human reason cannot ascertain were they not revealed in scripture), Ferguson made a number of points intended to defend his own Calvinist doctrinal views by dividing "pure" revelations into two subcategories. In the first subcategory he included those revelations that he claimed human reason could understand and defend once they had been revealed, such as the resurrection of the body and Christ's Satisfaction.[87] Where these mysteries are concerned, Christians could explain

85. Robert Ferguson, *The Interest of Reason in Religion with the Import & Use of Scripture-Metaphors; and the Nature of the Union Betwixt Christ & Believers* (London, 1675), p. 7.
86. Ferguson, *Interest*, pp. 14–28, 240–241.
87. Ferguson, *Interest*, pp. 33–35; 224–232.

"how these things can be" either "from principles of Natural Light, or from the account that the Scripture it self Gives of the *Modes* of their Existence."[88]

The second subcategory, on the other hand, included those revelations (such as the Trinity and the Incarnation) that, even after they had been revealed, reason could neither comprehend nor demonstrate. Where these mysteries are concerned, Ferguson argued, Christians cannot understand "the manner and way how they exist."[89] Although reason could neither comprehend nor rationally defend these mysteries, reason could (1) show that they could not be comprehended; (2) judge any argument against these revelations to be a "sophism," even if the fallacy in such arguments could not be detected; and (3) answer the objection by showing that the objection being forwarded was based on principles (or axioms or maxims) that had been extended beyond their sphere, for "to urge them beyond their bounds, is to contradict Reason, which tells us that [such axioms] hold only so far, and no farther."[90]

In discussing this third point Ferguson made a connection between the Cartesian criterion of clear and distinct ideas and the arguments of the Socinians, with which he showed considerable familiarity.[91] Even the Socinians had admitted, Ferguson argued, that there are "many things in the Christian Religion, which are above . . . Reason" and knew that "Religion transcends reason."[92] Yet despite this admission, the Socinians rejected such revelations as the Trinity and the Incarnation because they considered those revelations to be actually repugnant to reason rather than simply beyond reason's sphere of competence. The problem, Ferguson argued, was that too many things were being urged as principles of reason (that is, as clear and distinct perceptions) that in fact were not clear and distinct at all because human reason had been impaired by Adam's fall. In short, although Ferguson believed depraved reason to be capable of comprehending and defending some Christian doctrines,

88. Ferguson, *Interest*, p. 225.
89. Ferguson, *Interest*, p. 226.
90. Ferguson, *Interest*, pp. 227–230; quote, p. 230.
91. Ferguson, *Interest*, pp. 252–274. Among others, Ferguson cited Ostorodt's *Unterrichtung von den vornemsten Hauptpuncten der christlichen Religion* (pp. 271, 396), Schlictingius's controversy with Meisner (p. 271), the *Racovian Catechism* (p. 396), Smalcius's *De divinitate Jesu Christi* (pp. 271–272; 396), Faustus Socinus's *De Jesu Christo servatore* (pp. 271, 396), Crellius's *Ad librum Hugonis Grotii* (p. 396), and Volkelius's *De vera religione* (p. 396).
92. Ferguson, *Interest*, p. 267.

other doctrines remain utterly mysterious and must be held as articles of faith.

Ferguson devoted the rest of *Interest* to rebutting the claim of the Anglican divine William Sherlock (1641–1707) that in remaining separate from the Church of England, dissenters had ipso facto separated themselves from Christ.[93] It is here that the polemical intent in his discussion of reason's competence and limits becomes clear, for in the course of his refutation of Sherlock's claim, Ferguson classified the truth that Christ's union with believers is an immediate union (and not one mediated by the Church) as being one of the mysterious articles of faith that reason could neither comprehend nor demonstrate. If Sherlock could accept other mysterious unions, such as that between minute particles of matter and the union between the immaterial soul and the human body, Ferguson argued, then he was not justified in claiming that Christ's union could not be immediate on the grounds that such a union was too mysterious to be believed. Although Ferguson readily admitted that he himself could not describe the exact nature of Christ's union with individual believers, he declared that there was no need to do so, for "there are Mysterious Doctrines in the Gospel" and "our Union with Christ is of that Number."[94]

Ferguson's strategy in *Interest* was complex and clever. Glanvill had accused him of denigrating the role of reason in religion, yet Ferguson had managed to extend reason's role even further into revelation than had Glanvill (by arguing that reason can comprehend and defend even some "pure" revelations). At the same time, he had responded to Sherlock's charge that in asserting the immediacy of Christ's union with individual believers he was asserting an incomprehensible mystery (by arguing that *some* "pure" revelations were in fact incomprehensible mysteries due to reason's having become depraved as a result of Adam's fall). In short, he was able to have his cake and eat it too; he was able to

---

93. Sherlock's target was John Owen's *Of Communion with God,* which had been published almost twenty years earlier. Owen immediately responded with *A Vindication of some Passages in a Discourse concerning Communion with God from the Exceptions of William Sherlock* (1674). In 1675, the same year that Ferguson entered the controversy on Owen's side, Sherlock responded to Owen's *Vindication* in *A Defence and Continuation of the Discourse concerning the Knowledge of Jesus Christ.* For a discussion of this controversy, see Dewey D. Wallace, Jr., *Puritans and Predestination: Grace in English Protestant Theology, 1525–1695* (Chapel Hill: University of North Carolina Press, 1982), pp. 170–173.

94. Ferguson, *Interest,* pp. 482–487; quote, p. 585.

appear even more rational than Glanvill, while at the same time he was able to shield his own interpretation of the nature of Christ's union with believers from the rational scrutiny of Anglican opponents such as Sherlock.

John Owen's Nature of Apostasie

To high Calvinists such as Owen, the moderate Anglicanism espoused by the clergymen of the Restoration Church of England seemed to be a repudiation of all that was true and holy; to them, the Church seemed to be straying from orthodoxy and teetering dangerously on the brink of heresy. In *The Nature of Apostasie* (1676), Owen laid part of the blame for this "Decay of the Power of Religion in the World," or "present general Defection from the Truth, Holiness, and Worship of the GOS- PEL" (the phrases come from his subtitle) on the Socinian insistence that scripture must be interpreted in such a way as to be intelligible to and consonant with human understanding.

Owen did not, in *Apostasie*, refer to his earlier *Vindiciæ* (1655), in which he himself had assumed reason's competence to defeat the doc- trines of the Socinians. Although he did note briefly in *Apostasie* that "we ought to imploy the utmost of our Rational Abilities in the Investi- gation of Sacred Truth,"[95] his emphasis had shifted in the intervening years. By 1676, Owen's emphasis was on reason's limits, not reason's competence; perhaps in this, his views had been influenced by his assis- tant, Robert Ferguson.

In *Apostasie*, Owen argued that although the Protestants were right to emancipate "Reason from under the bond of Superstition & Tradition" in which it had been held by the Roman Catholics, the influence of the Socinians in recent years had caused reason to "enlarge its Teritories."[96] In describing this "enlargement," he complained that the Socinians had been so successful in introducing their doctrines because they first argued

---

95. John Owen, *The Nature of Apostasie from the Profession of the Gospel, and the punishment of apostates declared, in an exposition of Heb. vi. 4– 6, etc.* (London, 1676), p. 281. The passage from Hebrews reads: "For it is impossible for those who were once enlightened, and have tasted of the heavenly gift, and were made partakers of the Holy Ghost, and have tasted the good word of God, and the powers of the world to come, if they shall fall away, to renew them again unto repentence; seeing they crucify to themselves the Son of God afresh, and put him to an open shame."

96. Owen, *Apostasie*, pp. 7–8.

that there is no Reason why we should believe any thing that Reason cannot comprehend, so that we may safely conclude, that whatever is above our Reason is contrary unto it, and for what is so, it is destructive to the very natural Constitution of our Souls not to reject. And secondly [they argued] that the mind of Man is in its present condition every way sufficient unto the whole of its duties both Intellectual and Moral, with respect unto God.[97]

Owen was not surprised to discover that having thus laid the groundwork by convincing others that reason is the "just measure and standard of Truth," the Socinians had been so successful in introducing their particular interpretations of the doctrines of Christ's Satisfaction and Atonement and the justification and sanctification of believers. Further, Owen complained, if one responded to the Socinians by arguing

> that there are some things in Religion that are above Reason, as it is finite and limited, and some things contrary unto it, as it is deprav'd and corrupted, . . . they will reply (what is true in it self but wofully abus'd) that yet their reason is the best, yea only means, which they have to judg of what is true or false.[98]

Unfortunately, Owen continued, "it is a part of the Depravation of our Nature not to discover its own Depravations."[99]

According to Owen, then, some articles of Christianity are not only *above* reason, they are also *"contrary unto Reason as corrupted."* And among the articles contrary to reason Owen included those concerning God's eternal decrees, the mediation of Christ and justification by means of his righteousness, and the role of grace in the conversion of sinners.[100]

Conclusions

Certain things should be noted in the responses to Socinianism examined here. Glanvill's *Seasonable Recommendation* is the only latitudinarian response in the group, and it is clear that in his view, human reason is competent to judge the content of revelation. Not all of his Anglican contemporaries agreed with him, however. Edward Stillingfleet (1635–1699), for example, argued reason's incompetence to comprehend fully the content of revelation in *Origines Sacræ* (1662) along much the same

97. Owen, *Apostasie*, p. 10.
98. Owen, *Apostasie*, p. 9.
99. Owen, *Apostasie*, p. 11.
100. Owen, *Apostasie*, p. 297.

lines as the nonconformists I have discussed.[101] In this work, Stillingfleet,
then rector at Sutton and later to be Bishop of Worcester, argued that the
principles of reason should not be extended to *"those things* [Christian
mysteries] . . . which were never *looked* at in the *forming* of them,"[102]
just as Ferguson would later argue that the principles of reason hold
"only so far, and no farther," and just as Boyle would argue, even later,
in *A Discourse of Things above Reason* (see chapter 4). For the most
part, however, conformist divines tended to avoid the question of
whether human reason is competent to *judge* the content of revelation,
emphasizing instead that reason provides grounds for *believing* revealed
doctrines, thereby often managing to dodge the issue of the correct
interpretation of those doctrines.[103]

The nonconformists, on the other hand, seem to have been relatively
consistent in emphasizing reason's incompetence to judge the content of
revelation, and it would be *these* arguments (along with Stillingfleet's)
that Boyle would ultimately use in *Things above Reason.* Boyle was
influenced more by the nonconformist tradition than previous scholars
have recognized.

It is also important to note that in the course of these three decades,
high Calvinists such as Owen and Ferguson came to use arguments
about reason's limits, which had previously been used to confute the
claims of the Socinians, to defend their own interpretations of doctrine
from the rational scrutiny of their Anglican opponents. Ferguson, for
example, shielded the "mystery" of the nature of Christ's union with the

101. For discussions of Stillingfleet's theology, see Richard H. Popkin, "The
Philosophy of Bishop Stillingfleet," *History of Philosophy* 9 (1971), pp.
303–319; Sarah Hutton, "Science, Philosophy, and Atheism: Edward Stillingfleet's Defence of Religion," in *Scepticism and Irreligion in the Seventeenth and Eighteenth Centuries,* edited by Richard H. Popkin and Arjo
Vanderjagt (Leiden: E.J. Brill, 1993), pp. 102–120. For Stillingfleet's relationship to Boyle, see Michael Hunter, "Casuistry in Action: Robert
Boyle's Confessional Interviews with Gilbert Burnet and Edward Stillingfleet, 1691," *Journal of Ecclesiastical History* 44 (1993), pp. 80–98.
102. Edward Stillingfleet, *Origines Sacræ; or, A Rational Account of the
Grounds of Christian Faith, as to the Truth and Divine Authority of the
Scriptures, and the matters therein contained* (London, 1662), p. 236.
103. Gerard Reedy, S.J., *The Bible and Reason: Anglicans and Scripture in
Late Seventeenth-Century England* (Philadelphia: University of Pennsylvania Press, 1985); John Spurr, *The Restoration Church of England,
1646–1689* (New Haven: Yale University Press, 1991), esp. pp. 250–
257; idem, " 'Rational Religion' in Restoration England," *Journal of the
History of Ideas* 49 (1988), pp. 563–585.

individual believer from rational scrutiny by claiming that it must be believed (because, or so Ferguson claimed, it was revealed), while at the same time acknowledging that he could not understand *how* it could be true; that was what made it a mystery. Similarly, Owen invoked reason's limits to shield the high Calvinist interpretation of predestination from rational scrutiny. Divines such as Ferguson and Owen were in effect claiming that they had reached a uniquely correct interpretation of that which they themselves had acknowledged was above (and sometimes even contrary to) human reason. And they were among the most polemical of the participants in the apparently endless controversies over the correct interpretation of the doctrine of predestination in seventeenth-century England.

In the next chapter, I examine the controversies over predestination, concluding with a discussion of John Howe's *Reconcileableness of God's Prescience of the Sins of Men, with the Wisdom and Sincerity of his Counsels, Exhortations, and whatsoever Means He uses to prevent them* (1677), an attempt to reconcile the high Calvinist and Arminian interpretations of predestination, and a work which he wrote at Boyle's request. Howe's *Reconcileableness,* instead of putting an end to controversy, generated yet another dispute over the question of predestination, and it was this dispute that prompted Boyle's *Things above Reason* (1681). In *Things above Reason,* Boyle would borrow and refurbish the arguments of the nonconformists. Boyle, however, would do so not with the intent to shield particular doctrinal interpretations from rational scrutiny, but with the intent to convince his contemporaries that it simply is not consistent to argue dogmatically about things that lie outside the competence of human understanding.

3

Predestination Controversies

Theological controversies in seventeenth-century England show clearly the impact of the proliferating and often contradictory interpretations of scripture that accompanied the development of Protestantism. And one of the most passionate controversies of all was the apparently endless one over the correct interpretation of predestination. As a modern commentator has aptly observed, "the Calvinist-Arminian controversy placed great strain on Luther's doctrine of the perspicuity of Scripture."[1] With the fate of his or her soul at stake, it was crucially important that each individual discover which of the conflicting interpretations was uniquely correct, for surely God knew which was true and surely God would punish those who erred. Further, the fate of the nation itself was at stake, for surely God's pleasure or displeasure would affect the course of national and international events. Hence it was crucially important that the divines of the Church of England determine correct doctrine and establish styles of worship that would be pleasing to God.

This was not an easy task. In order to unravel the *"great mystery of godliness"* (or "the great work of man's redemption"), Boyle himself noted that

> a man must know much of the nature of spirits in general, and even of the father of them God himself; of the intellect, will, &c. of the soul of man; of the state of *Adam* in paradise, and after his fall; of the influence of his fall upon his posterity, of the natural or arbitrary vindictive justice of God; of the grounds and ends of God's inflicting punishments, as a creditor, a ruler, or both; of the admirable and

---

1. Alan C. Clifford, *Atonement and Justification: English Evangelical Theology 1640–1790. An Evaluation* (Oxford: Clarendon Press, 1990), p. 142. As the word "evaluation" in the title suggests, Clifford is concerned to reach a correct verdict on the controversy between Arminians and Calvinists; nevertheless, his detailed scholarship on the arguments of three of the seventeenth-century controversialists (John Owen, Richard Baxter, and John Tillotson) is excellent.

unparalled person of Christ the mediator; of those qualifications and offices, that are required to fit him for being lapsed man's redeemer; of the nature of covenants, and the conditions of those God vouchsafed to make with man, whether of works, or grace; of the divine decrees, in reference to man's final state; of the secret and powerful operations of grace upon the mind, and the manner, by which the spirit of God works upon the souls of men, that he converts, and brings by sanctification to glory: to be short, there are so many points (for I have left divers unamed) most of them of difficult speculation, that are fit to be discussed by him, that would solidly and fully treat of the world's redemption by *Jesus Christ*, that when I reflect on them, I am ready to exclaim with St. *Paul*, *who is sufficient for these things?*[2]

Unfortunately, in Boyle's view, many divines felt that they were "sufficient for these things." Equally unfortunately, they did not agree about "these things."

Arminians versus Calvinists

Scholars disagree as to how the controversies between Arminians and Calvinists over the salvation process in seventeenth-century England are best interpreted.[3] Some scholars argue that the two camps were in essential agreement, that their polemics involved differences in emphasis rather than substantial differences, and that the actual positions held fall along a continuum rather than being easily categorized into two opposing positions.[4] Certainly there is considerable truth in this claim. The English Protestants were, after all, just that: Protestants. As such, they shared the fundamental beliefs of Protestantism, and when writing against their perceived common enemies, the Roman Catholics and the atheists, they tended to minimize their differences.[5]

---

2. Boyle, *High Veneration, Works*, v. 5, p. 144.
3. An excellent historiographical survey may be found in J. Sears McGee, Introduction to *The Godly Man in Stuart England: Anglicans, Puritans, and the Two Tables, 1620–1670* (New Haven, CT: Yale University Press, 1976). More recent studies are cited in the subsequent notes of this chapter.
4. Peter White, in *Predestination, Policy and Polemic: Conflict and Consensus in the English Church from the Reformation to the Civil War* (Cambridge: Cambridge University Press, 1992), stresses the existence of a spectrum of beliefs rather than a polarity.
5. For the perceived threat of atheism in seventeenth-century England see

Other scholars, however, looking not so much at the objective content of the differences between Anglican and Puritan but more at the way those differences were perceived subjectively by the contemporary controversialists themselves, argue that to the seventeenth-century polemicists, the differences in interpretation certainly seemed substantive.[6] This, in my view, is the more accurate interpretation. Although there were at the time some irenical souls, moderates on both sides, who believed that the two positions could be reconciled if only extremists on both sides of the spectrum would bend just a little, the unyielding tenacity with which many clung to their positions is an indication of their perception of the significance of the differences between opposing views.

Michael Hunter, "Science and Heterodoxy: An Early Modern Problem Reconsidered," in *Reappraisals of the Scientific Revolution*, edited by D.C. Lindberg and R.S. Westman (Cambridge: Cambridge University Press, 1990), pp. 437–460; idem, " 'Aikenhead the Atheist': The Context and Consequences of Articulate Irreligion in the Late Seventeenth Century," in *Atheism from the Reformation to the Enlightenment*, edited by Michael Hunter and David Wootton (Oxford: Clarendon Press, 1992), pp. 221–254; Nigel Smith, "The Charge of Atheism and the Language of Radical Speculation, 1640–1660," in *Atheism from the Reformation to the Enlightenment*, pp. 131–158. For anti-Catholicism, see Peter Lake, "Anti-popery: the Structure of a Prejudice," in *Conflict in Early Stuart England; Studies in Religion and Politics 1603–1642*, edited by Richard Cust and Ann Hughes (London and New York: Longman, 1989), pp. 72–106; Henry G. van Leeuwen, *The Problem of Certainty in English Thought, 1630–1690* (The Hague: Martinus Nijhoff, 1963), chapter 2. See also Raymond D. Tumbleson, " 'Reason and Religion': The Science of Anglicanism," *Journal of the History of Ideas* 57 (1996), pp. 131–156. Tumbleson's conclusions about Boyle are founded on the mistaken assumption that Boyle was the author of "Reasons Why a Protestant Should not Turn Papist; or, Protestant Prejudices against the Roman Catholick Religion." On the authorship of this essay, see Edward B. Davis, "The Anonymous Works of Robert Boyle and the *Reasons Why a Protestant Should not Turn Papist* (1687)," *Journal of the History of Ideas* 55 (1994), pp. 611–29.

6. See, for example, Isabel Rivers, *Reason, Grace, and Sentiment: A Study of the Language of Religion and Ethics in England, 1660–1780*, vol. 1, *Whichcote to Wesley* (Cambridge: Cambridge University Press, 1991); Dewey D. Wallace, Jr., *Puritans and Predestination, Grace in English Protestant Theology, 1525–1695* (Chapel Hill: University of North Carolina Press, 1982), J. Sears McGee, *The Godly Man in Stuart England*; and John F. H. New, *Anglican and Puritan: The Basis of their Opposition, 1558–1640* (Stanford, CA: Stanford University Press, 1964).

Doctrinal Issues

The doctrinal issues involved in the correct interpretation of predestination are complex, as Boyle noted in the passage quoted above. The heart of the problem is two apparently conflicting views on the subject expressed in the *New Testament:* Romans, chapter 3, verse 28, and James, chapter 2, verse 24. Calvinists invoked the passage from Romans ("Therefore we conclude that a man is justified by faith without the deeds of the law"), while Arminians appealed to the passage in James ("Ye see then how that by works a man is justified, and not by faith only"). As one scholar has noted, this "seeming discrepancy between the apostles Paul and James ... has been a perpetual source of embarrassment since the time of the Reformation."[7]

The differences between these two views can be approached in a number of different ways: as a question of faith versus works, as a question of justification, as a question concerning God's attributes, or as a sacramental issue. Further, they can be examined not only as doctrines actually held but as they were perceived to be held, for contemporaries often conflated the position of opponents with the most egregious of heresies. Readers interested in a comprehensive treatment of the topic must look elsewhere.[8] My purpose here is to outline the different views in sufficient detail to enable readers unfamiliar with these seventeenth-century controversies to understand the significance of the issues relevant to this book without going into fine details peripheral to the concerns of this study, and my emphasis is on those aspects of the controversies that were of particular concern to Boyle.

John Humphrey's Account

Perhaps the best way to approach the subject is to examine the various possible positions and the consequences of those positions as seen by a contemporary, John Humphrey, who summarized the issues involved in such a way (or so he claimed) as to be "Doing nothing by Partiality," as commanded in I Timothy, chapter 5, verse 21.[9] In *The Middle-Way in*

7. Clifford, *Atonement and Justification,* p. 221. For the Reformation debate, see *Erasmus-Luther Discourse on Free Will,* translated and edited by Ernst F. Winter (New York: Frederick Ungar Publishing Co., 1974).
8. See, for example, the works cited in notes 1, 3, 4 and 6.
9. The verse is, "I charge *thee* before God, and the Lord Jesus Christ, and the elect angels, that thou observe these things without preferring one before another, doing nothing by partiality." Humphrey's point was that both

One Paper of Election & Redemption, with Indifferency between the Arminian & Calvinist (1673), Humphrey, a nonconformist who promoted bills for comprehension and toleration among members of Parliament,[10] began by noting that both Arminian and Calvinist divines agreed that God has foreknowledge of human affairs. With that in mind, he went on to say that election can be interpreted in two general ways, either as absolute or as conditional. In absolute predestination, the Calvinist position,[11] God, in decreeing who shall be saved, does not consider either works or faith, but simply decides (without informing human beings of his reasons) to grant his saving grace to some individuals and to withhold his grace from others. Absolute predestination can itself be interpreted in two ways. According to the supralapsarian doctrine, God's decrees were determined prior to Adam's fall and are unrelated to the subsequent state of sin, whereas according to the sublapsarian doctrine, God's decrees are based on the human condition after the fall.

In conditional predestination, on the other hand, God's decrees are based on his foreknowledge of human affairs. Conditional predestination itself can be interpreted in two different ways. The first is that God's decrees are based on his foreknowledge of works (that is, of human conduct). This, as Humphrey correctly pointed out, is the Pelagian position, and was considered to be heretical by orthodox Roman Catholics and Protestants alike. In the second view, the Arminian position, God's decrees are based on his foreknowledge of who will believe; this, unlike the Pelagian "election of works," is the "election of faith."[12]

Calvinists and Arminians placed undue emphasis on different (and apparently conflicting) "preferred" passages of scripture.

10. For a brief discussion of Humphrey's efforts, see John Spurr, *The Restoration Church of England, 1646–1689* (New Haven, CT: Yale University Press, 1991), pp. 70–71.

11. For studies of the transformation of Calvin's thought to Reformed orthodoxy, see Brian G. Armstrong, *Calvinism and the Amyraut Heresy: Protestant Scholasticism and Humanism in Seventeenth-Century France* (Madison: The University of Wisconsin Press, 1969), and Richard A. Muller, *Christ and the Decree: Christology and Predestination in Reformed Theology from Calvin to Perkins* (Durham, NC: The Labyrinth Press, 1986).

12. J[ohn] H[umphrey], *The Middle-Way in one Paper of Election & Redemption, with Indifferency between the Arminian & Calvinist* (London, 1673), pp. 1–5. Humphrey's solution is to distinguish between power and will, arguing that all human beings have the power to believe, but that not everyone who will actually believe has the *will* to do so. Hence God gives some individuals the will to believe; why God has chosen to grant that gift to some and not to others is incomprehensible to human beings (p. 9).

Corollaries and Consequences

A number of corollaries to and consequences of these two positions should be noted, although individuals on both sides offered nuances of interpretation and emphasis that make such generalizations hazardous.[13] According to the Calvinists, Christ died in order to impute his righteousness to those whom God had elected to eternal life, whereas according to the Arminians, Christ died for all sinners, even though some (or even most) individuals might reject this freely offered gift of salvation. The Calvinists' emphasis was on God's absolute power; the Arminians, on the other hand, emphasized God's goodness and mercy. The Calvinists stressed man's impotence, whereas the Arminians stressed man's free will and power to accept and cooperate with or to reject God's proffered grace.

Individuals on each side saw the opposing position as involving or implying impious (if not heretical) consequences. Arminians argued that a Calvinist God would have created some souls simply in order to damn them, which would contradict God's goodness and make God, in effect, the author of sin. Arminians also pointed out that a wise and sincere God would not have filled scripture with his exhortations to human beings not to sin if individuals did not possess the will and power to obey those exhortations. Further, the Calvinist view of the inevitability of the salvation of the elect led to charges (rarely warranted) that Calvinists were antinomians (those who believed that grace released the believer from the obligation to obey the moral law).[14] Calvinists, on the other hand, charged the Arminians with Papism, Pelagianism, and Socinianism, thus conveniently lumping together these radically different groups of Christians simply on the basis of their common doctrine of conditional predestination. Further, Calvinists accused the Arminians of being "mere moralists" with their emphasis on the role of human cooperation with God's grace, and (perhaps worst of all) of devaluing God's omnipotence by making his decrees dependent on human contingencies.

13. On the hazards of such generalizations, as well as their justification in the context of the controversies between Arminians and Calvinists, see McGee, *Godly Men,* pp. 1–2. For a different assessment on the validity of such generalizations, see Peter White, *Predestination, Policy and Polemic,* pp. xii-xiii.
14. For differing conceptions of antinomianism in seventeenth-century England, see Spurr, *Restoration Church of England,* pp. 304-305.

Law of Noncontradiction

Such a brief sketch does not do justice to the complexities of the two opposing views, and before turning first to Boyle's early treatments of the conflicts between the Calvinists and the Arminians and then to the specific controversy that occasioned his *Discourse of Things above Reason,* I will stress one aspect of those complexities that is not always immediately apparent but was always lurking in the background. This aspect is the role of the law of noncontradiction in the controversies, for both the Calvinists and Arminians were perceived by their opponents as believing in absurdities and impossibilities. "Which side soever of the question you take," Boyle noted, "you will be unable directly and truly to answer the objections that may be urged to shew, that you contradict some primitive or some other acknowledged truth."[15] Although he made this comment in reference to a different controversy – one concerning the endless divisibility of a straight line – it is indeed relevant to the controversies over the proper interpretation of predestination. Arminians and Calvinists alike believed that God is omnipotent, omniscient, and infinitely good. Hence Calvinists could claim that the Arminian emphasis on free will, insofar as it attributed some power in regard to salvation to the individual, was inconsistent with God's omnipotence. And the Arminians could claim that the Calvinist emphasis on reprobation (or God's having created some souls knowing that they would be damned) made God the author of sin, and hence not infinitely good. Further, Arminians could charge that if God's will concerning election and reprobation is absolute, then God's wisdom in exhorting individuals not to sin is impeached. And the difficulty (if not impossibility) of reconciling God's prescience with the free will of human beings has posed difficulties for philosophers throughout the ages; if both cannot be affirmed, then one, at least, must be denied.[16]

Boyle's *Seraphic Love*

Boyle included an extended discussion of the difficulties involved in the conflicts between Arminian and Calvinist views in the first work published under his own name, *Some Motives and Incentives to the Love of God* (1659), most often cited as *Seraphic Love* (its running title). In this

---

15. Boyle, *Things above Reason, Works,* vol. 4, p. 423.
16. The most comprehensive study of this problem is William Lane Craig, *The Problem of Divine Foreknowledge and Future Contingents from Aristotle to Suarez* (Leiden: E.J. Brill, 1988).

work, he indicated clearly his belief that difficulties involved in accurately interpreting God's revelation concerning the salvation process were due at least partially to the limits of human understanding.

Boyle's purpose in *Seraphic Love* was, as he put it, to foster in human beings the same "burning love" for God that God has for human beings. In the process, he emphasized that God's love for human beings springs from his own nature and not from anything that humans might initiate. Realizing that his comments might be interpreted as lending public support to the Calvinist interpretation of predestination, he noted that his comments might initially sound as if they presupposed the "truth of their doctrine, who ascribe to God, in relation to every man, an eternal, unchangeable and inconditionate decree of election, or reprobation."[17] Although refusing to state his own "sense" of the "controversies betwixt the Calvinists and the Remonstrants [i.e., Arminians]," he went on to emphasize that because the Calvinist doctrine of predestination

> is not only by almost all the rest of mankind, but by the rest of the protestant churches themselves (the Lutherans, and diverse learned divines of the church of *England*) not only rejected, but detested as little less than blasphemous (as indeed they, that judge it an error, cannot but be tempted to think it a dangerous one, and of very pernicious consequence, so far forth as its sequels are permitted to have influence on men's practice) I think it not amiss to advertise you, that the doctrine of [absolute] predestination is not necessary to justify the freeness and the greatness of God's love.[18]

He went on to declare that he did not consider the two positions to be irreconcilable:

> Those, that are truly pious of either party, are perhaps otherwise looked on by God than by any other, as contending, which of God's attributes should be most respected; the one seeming to affirm irrespective decrees, to magnify his goodness; and the other to deny them, but to secure the credit of his justice. And even in honouring the same attribute, his goodness, these adversaries seem rivals, the one party supposing it best celebrated by believing it so irresistible, that to whomsoever it is intended, he cannot but be happy [i.e., saved]; and the other thinking it most extolled by being so universal, that it will make every man happy, if he pleases.[19]

Boyle continued by arguing that the dispute between the Calvinists and Arminians was "not so much concerning the thing, as concerning

---

17. Boyle, *Seraphic Love, Works*, vol. 1, p. 277.
18. Boyle, *Seraphic Love, Works*, vol. 1, pp. 277–278.
19. Boyle, *Seraphic Love, Works*, vol. 1, p. 277. The "pernicious effects" Boyle noted as a possible result of Calvinism is most likely a reference to antinomianism.

the manner of its being proffered," the Calvinists affirming God's grace to be "irresistibly presented," and the Arminians, although denying that grace is irresistible, affirming along with the Calvinists that grace is altogether free and undeserved and that it is also offered along with a power enabling those to whom it is offered to accept it. Indeed, he insisted, people are generally so conscious of their debt to God that even those who were mistaken in their interpretation of the salvation process – the Roman Catholics, the Socinians, and the Jews – acknowledged that indebtedness. "The more sober sort of Romish catholics," Boyle observed, ascribe more to God's grace and less to man's merit than "the more quarrelsome writers of their party" have led Protestants to believe. The Socinians, he continued, attempted to "free their erroneous doctrine of justification" from the charge that they had only themselves to thank for their salvation. Even the Jews, those "great patrons of man's free-will," acknowledged God's grace, he claimed, citing "Manasseh Ben-Israel," as well as a Jewish professor of Hebrew who had assisted him in his study of that language.[20]

Boyle's point was that at least the more "sober" and "fair" members of each of these groups acknowledged the freeness of God's grace, although the proponents of each of the various interpretations of the salvation process emphasized (some to the point of error) one or another aspect of God's plan. Boyle's conclusion was that while we are on earth, "the darkness that is here cast on all things" and "the dimness of our intellectual eyes" allow humans to discern but little of God's "wisdom, power, and goodness." In heaven, human understanding will be enlarged; in heaven we will understand why some passages of scripture had to be so obscure.[21]

Interestingly, the passages from *Seraphic Love* cited here are not present in an extant copy of the manuscript as written in 1648 or 1649 and only recently discovered by Lawrence M. Principe, but were added at

---

20. Boyle, *Seraphic Love, Works,* vol. 1, pp. 278–279.
21. Boyle, *Seraphic Love, Works,* vol. 1, pp. 278–279, 290, 283, 289. In the *Appendix* to *Christian Virtuoso* (*Works,* vol. 6, p. 694) Boyle noted that the limits of human understanding necessitate the contemplation of God's attributes piecemeal: "Those different conceptions of a divine object, whose nature is most simple, flow from the incapacity of our limited and imperfect understandings, to frame a conception that shall comprehend the infinitely perfect nature of God in one single and simple idea; and therefore we are reduced to consider and represent him as it were in parts; contemplating him sometimes as omnipotent, and sometimes as wise, and sometimes as just."

some time between those years and the publication of *Seraphic Love* in 1659.[22] As I showed in chapter 2, Socinianism, with its insistence that scripture must be interpreted in such a way as to be consistent with human reason, was growing in influence in England during the 1650s. Boyle's concern about this (in his opinion) misguided insistence on the competence of human reason was expressed in his "Essay of the holy Scripture" (discussed in chapter 2), as well as in the passages I have cited from *Seraphic Love*.

It is difficult to assess Boyle's own "sense" of the controversy from his comments in *Seraphic Love*. Despite having noted that the doctrine of predestination (interpreted along Calvinist lines) was not necessary to justify God's grace, he listed Arminians along with the Roman Catholics, Socinians, and Jews as among those who held mistaken views, and the tone of *Seraphic Love* as a whole is certainly more Calvinistic than Arminian. Perhaps it is significant that these passages were written during the Interregnum, when Calvinism was at its height. Later, Boyle tended to place at least equal emphasis on the role of free will in the salvation process. In *The Reconcileableness of Reason and Religion* (1675), for example, he again noted the difficulties involved in reconciling the two views. There he noted that some Socinians denied that man's free will is the proper object of foreknowledge, and commented that even the Remonstrants (or Arminians), who affirmed both prescience and free will, did not know how to reconcile them. Nevertheless, he claimed, Christians are compelled to acknowledge prescience by the infiniteness ascribed to God's perfections as well as by prophetic predictions and free will because it is "evident."[23]

Against this background, I now turn to the specific controversy over predestination that led to the publication in 1681 of Boyle's *Discourse of Things above Reason* and its accompanying *Advices in judging of Things said to transcend Reason*.

Howe's *Reconcileableness* and Hammond's *Pacifick Discourse*

John Howe was an irenical soul who attempted to reconcile the Calvinist and Arminian positions on predestination in his *Reconcileableness of*

22. Lawrence M. Principe, "Style and Thought of the Early Boyle: Discovery of the 1648 Manuscript of *Seraphic Love*," *Isis* 85 (1994), pp. 247–260.
23. Boyle, *Reason and Religion, Works*, vol. 4, p. 176; *Advices, Works*, vol. 4, p. 466.

God's Prescience of the Sins of Men, with the Wisdom and Sincerity of his Counsels, Exhortations, and whatsoever Means He uses to prevent them (1677), which he had written, he said, "at the request of Mr. Boyle," and which took the form of "A Letter to the Honourable Robert Boyle Esq."[24]. A Presbyterian, Howe had refused to conform to the Church of England in 1662, partly on the grounds that submitting to reordination within the Church would constitute a denial of the validity of his Interregnum ministry, and partly on the grounds that he opposed forced conformity.[25] From 1662 to 1671, he conducted private worship services, and from 1671 to 1675, he served as a personal chaplain in Ireland. In 1675, he became co-pastor of the Presbyterian congregation in Cheapside (London) and therefore would have been in London at the time that he and Boyle might have been discussing the writing of *Reconcileableness*.

The exact nature of Howe's relationship to Boyle is not known. There does not seem to be any surviving record of correspondence between the two men. I have been unable to find any such correspondence among the Boyle Papers and Letters at the Royal Society in London, and Howe's papers were destroyed, at his request, shortly before his death.[26] There is, however, evidence of communication between the two men in two of Joseph Glanvill's letters to Boyle.[27] Another link between the two might

---

24. The full title of Howe's work is *The Reconcileableness of God's Prescience of the Sins of Men, with the Wisdom and Sincerity of his counsels, Exhortations, and whatsoever Means He uses to prevent them, in a Letter to the Honourable Robert Boyle Esq.* (London, 1677); this work was originally published under Howe's "penultimate letters" ("H.W."). My citations come from the edition of the same year, *To which is added a Post-Script in London 1677, Defence of the said Letter*; the *Post-Script* is separately paginated. Because the original publication of *Reconcileableness* precipitated a controversy, Howe considered it more "becoming" for the subsequent edition to be published under his full name (*Postscript*, pp. 2–3). Unless otherwise noted, my citations to this work refer to the body of the original text rather than to the *Postscript*. Howe's statement that he wrote the work at Boyle's request is in the Dedicatory Epistle.

25. Henry Rogers, *The Life and Character of John Howe, M.A., with an Analysis of his Writings* (London, 1836), pp. 134, 152.

26. Rogers, *Life and Character of John Howe*, p. 2.

27. Glanvill to Boyle, letters of Oct. 7 and Feb. 24, *Works*, vol. 6, pp. 631–632 and 632–633. In the first of these letters, Glanvill states that he has heard from "Mr. *Jo. How*, the minister," that Howe had informed Boyle of Glanvill's plan to collect evidence of the existence of witches, and that Boyle had responded by noting an account he would send to Glanvill; a

have been Edward Stillingfleet. Howe's biographer claims that Howe was on intimate terms with Stillingfleet, one of Boyle's confessors.[28]

The Immediate Background

I do not know whether any specific work prompted Boyle to ask Howe to write *Reconcileableness*. There were, however, any number of works in the 1650s, 1660s, and 1670s in which themes relevant to Howe's *Reconcileableness* were discussed and that might have caught both Boyle's and Howe's attention. For example, predestination was one of the doctrinal issues that the Socinian Joachim Stegmann considered in *Brevis Disquisitio: Or, a Brief Enquiry Touching a Better way Then is commonly made use of, to refute Papists, and reduce Protestants to certainty and Unity in Religion*, translated by John Biddle in 1653.[29] In that work, Stegmann asked whether it is possible for human beings to obey Christ's commandments. In answering in the affirmative, Stegmann argued against original sin on the grounds that such a doctrine strips humans of all power to obey God. Where predestination is concerned, he claimed that the doctrine means only that before the foundation of the world, God had determined to save all who would obey his commandments. This doctrine has been misinterpreted, Stegmann argued, by those who interpret predestination to mean that God has foreknowledge of which individuals will have faith because those Christians also ascribe that faith to the power of God, not man. This, he claimed, is wrong, for it lays "all the blame of sin and damnation" on God, and is contrary to God's wisdom and justice. Such a doctrine makes God like Tiberius, who ordered the execution of a virgin, and who then, knowing that such an execution was unjust, had the hangman deflower her.[30] Besides, he continued, what is the purpose of God's

---

reference to John Webster's denial of the existence of witches in *The Displaying of Supposed Witchcraft* (London, 1677) provides a clue to the year. In the second letter, Glanvill remarks that he is looking forward to receiving an account of witchcraft from Boyle via Howe.

28. Rogers, *Life and Character of John Howe*, p. 215; Michael Hunter, "Casuistry in Action: Robert Boyle's Confessional Interviews with Gilbert Burnet and Edward Stillingfleet, 1691," *Journal of Ecclesiastical History* 44 (1993), pp. 80–98.

29. This work is also discussed in chapter 2.

30. Stegmann's point was that if salvation is totally dependent on God's grace, and if God withholds his grace where some souls are concerned (resulting in the damnation of those souls), then the responsibility for damned souls

exhortations to men not to sin if man does not have the free will necessary to obedience?[31]

Predestination was similarly discussed in the *Racovian Catechism*.[32] There, both it (and the related doctrine of original sin) were denied on the grounds (among others) that they are contrary to God's nature, for God "is the fountain of all equity."[33] Adam's sin was but one act and could hardly have altered his own essential nature, much less that of all his posterity; references in scripture invoked in support of the doctrine of original sin are, according to the authors of the *Catechism*, in fact references to voluntary sin. And predestination, if true, would charge God with "many absurd, yea, horrid things." It would make God unjust, for example, if he were to punish the impious and disobedient for that which they cannot avoid doing, and it would make God a hypocrite and deceitful if he has already excluded some from the salvation he offers to all. Further, it would make God the author of sin, because "since it is altogether necessary that sin should go before damnation, certainly he that decrees a man shall be of necessity damned, decrees also that he shal of necessity sin."[34]

I noted in chapter 2 that some Socinians, in denying the doctrines of

---

falls on God (who is analogous to Tiberius and the hangman) rather than on human beings.

31. Joachim Stegmann, *Brevis Disquisitio: Or, a Brief Enquiry Touching a Better way Then is commonly made use of, to refute Papists, and reduce Protestants to certainty and Unity in Religion*, originally published 1633, translated by John Biddle (London, 1653), pp. 33–38; quote, p. 34.

32. *Catechism of the Church of those people who, in the Kingdom of Poland and the Grand Duchy of Lithuania and in other Domains belonging to the Crown, affirm and confess that no other than the Father of our Lord Jesus Christ is the only God of Israel and that the Man Jesus the Nazarene, who was born of the Virgin, and no other besides him, is the only begotten Son of God* (in Polish, Raków, 1605) was published by Biddle under the title of *The Racovian Catechisme: Wherein you have the substance of the Confession of those Churches, which in the Kingdom of Poland, and Great Dukedome of Lithuania, and other Provinces appertaining to that kingdom, do affirm, that no other save the Father of our Lord Jesus Christ, is that one God of Israel, and that the man Jesus of Nazareth, who was born of the Virgin, and no other besides, or before him, is the onely begotten Sonne of God* (Amsterledam [London], 1652). All subsequent references to the *Racovian Catechism* are to Biddle's edition. Other aspects of this work are discussed in chapter 2.

33. *Racovian Catechism*, p. 142.

34. *Racovian Catechism*, pp. 145–146.

predestination and original sin and in affirming man's free will, had explicitly denied God's prescience, and this was commented on with some regularity during the years preceding Howe's *Reconcileableness*.[35] Boyle himself had noted the Socinian denial of God's prescience in *The Reconcileableness of Reason and Religion* (1675), and had commented that although both prescience and free will must be affirmed, exactly how the two truths might be reconciled was not easy to discern.[36] And John Humphrey had denied in *Middle-Way* that God has foreknowledge of *all* future contingents, his point being that God could not foresee who might reject the gift of grace. As Humphrey expressed the problem, how could God be considered just if human beings do not have free will? How could God condemn any if he alone is responsible for the salvation of individuals? He went on to ask,

> Suppose a Magistrate for some fault shall cut off a Malefactors hands, and then command him to write, and promise him great things if he will, and threaten him if he do not write, can we think this Magistrate serious? Is this becoming a righteous and good man? If you grant not man to have power, why do you *Preach*? why do you *Exhort*? If you will give him more than this (I must persist), why do you *Pray*? why do you ask Grace of God? what is it to ask his Grace, but that God should incline the will?[37]

Alternatively, some authors denied that human beings possess free will rather than denying God's prescience. Peter Sterry, for example, who had served as a chaplain to Richard Cromwell and who hovered on the periphery of the Cambridge Platonists, denied free will on the grounds of the perfection of God's knowledge in his posthumously published *Discourse on the Freedom of the Will* (1675).[38] And certainly Hobbes's

---

35. See, for example, Henry Hammond, ΧΑΡΙΣ ΚΑΙ ΕΙΡΗΝΗ; *or, A Pacifick discourse of Gods Grace and Decrees: In a Letter, of full Accordance written to the Reverend, and most learned, Dr. Robert Sanderson, to which are annexed the Extracts of three Letters concerning Gods Prescience reconciled with Liberty and Contingency* (London, 1660), unpaginated preface, section 8 and pp. 101–102; Henry Hickman, *Historia Quinq-Articularis Exarticulata* (London, 1673), p. 17.

36. Boyle, *Reason and Religion, Works*, vol. 4, p. 176.

37. Humphrey, *Middle-Way*, pp. 14–20; quote, p. 17. The full title of Howe's work, published just two years after Humphrey's, reveals a possible connection between the two works: *The Reconcileableness of God's Prescience of the Sins of Men, with the Wisdom and Sincerity of his Counsels, Exhortations, and whatsoever Means He uses to prevent them.*

38. Peter Sterry, *A Discourse on the Freedom of the Will* (London, 1675), pp. 24–40.

and Spinoza's denial of free will was ever-present in the minds of those who believed freedom of the will to be essential to Christianity.

Further, some authors emphasized God's exhortations to human beings not to sin. In 1673, for example, the Arminian Samuel Hoard's *Gods Love to Mankind* (originally published in 1633) was republished in London; in this work, Hoard had rejected sublapsarian predestination on the grounds that God's predetermination that some individuals should perish is inconsistent with his exhortations to men not to sin.[39] In 1675, the Anglican divine Thomas Hotchkiss published *Reformation or Ruine,* in which he urged individuals to reform their ways and comply with God's exhortations not to sin, basing his work on Leviticus, chapter 26, verses 23–24, which read, "And if by this discipline you are not turned to me, but walk contrary to me, then I also will walk contrary to you, and I myself will smite you sevenfold for your sins."[40] Perhaps one or more of these many works prompted Boyle to encourage Howe to write his *Reconcileableness,* or perhaps some other consideration prompted the effort. What is clear is that controversies over predestination were an ongoing concern during the decades preceding Howe's effort, and that those who affirmed God's prescience left themselves open to the objection that God has urged that which he knows will not come to pass, and this was the particular issue that Howe addressed.

Howe's Argument

Howe's argument in *Reconcileableness* was not an original one; essentially, he followed the argument that had been presented by the Arminian Henry Hammond in 1660 to Robert Sanderson in Hammond's ΧΑΡΙΣ ΚΑΙ ΕΙΡΗΝΗ; *or, A Pacifick discourse of God's Grace and Decrees.* Sanderson, later to become Bishop of Lincoln, was an Anglican divine who had in the late 1650s revised his Calvinistic views. Hammond published his argument in the form of a letter to Sanderson as a way of making Sanderson's recent decision to affirm the role of free will in the salvation process known to their contemporaries and to reinforce Sanderson's new-found Arminianism. (It is perhaps relevant that about the time that Hammond published *Pacifick discourse,* Boyle was subsi-

39. Samuel Hoard, *Gods Love to Mankind, manifested by disproving his absolute decree for their damnation* (London, 1673), pp. 157–170.
40. Thomas Hotchkiss, *Reformation or Ruine: Being Certain Sermons upon Levit. XXVI.23, 24. First Preached, and afterwards with Necessary Enlargements fitted for Publick Use* (London, 1675).

dizing Sanderson's writings on cases of conscience, using Thomas Barlow, Bishop of Lincoln, as an intermediary.)[41]

Howe expressed the problem this way: If God foreknows that an individual will commit a particular sin, then in exhorting individuals not to commit that sin, God's wisdom is contradicted, "for we judg it not to consist with the wisdom of a man, to design and pursue an end, that which he foreknows he shall never attain." Further, God's sincerity ("his uprightness and Truth") is contradicted because his exhortations cannot be sincere if he knows that they will not be heeded. In short, "either the holy God seriously intends the prevention of such foreseen sinful actions and omissions or he doth not intend it. If he do, His wisdom seems liable to be impleaded. . . . If he do not, his uprightness and Truth."[42] According to Howe, however, God's prescience of the future actions of an individual simply amounts to God's knowing that the individual *will* do that which ought not be done; His foreknowledge in no way necessitates those actions.[43]

An essential aspect of both Howe's and Hammond's arguments was the attempt to reconcile the Arminian doctrine of conditional predestination with the Calvinist doctrine of absolute predestination by arguing that God may, but does not always, effect the salvation of an individual by overwhelming that individual with his grace. Thus God's grace is offered to all individuals (and this affirms God's wisdom in exhorting individuals not to sin), even though the salvation of some individuals who resist God's offer may be effected by a "super-effluency" of grace.[44]

Although this appeal to a "super-effluency" of grace accommodates the claims of both Arminians and Calvinists, it does not effectively reconcile the two positions. It simply raises the question of whether individuals have the power to resist God's grace to a different level. Instead of arguing whether individuals can resist God's grace (with Arminians arguing that they can and Calvinists arguing that they cannot), the opposing sides could debate instead whether individuals have the power to resist God's "super-effluent" grace. Both Hammond and

41. Correspondence between Boyle and Barlow on this subsidy may be found in *Works*, vol. 6, pp. 301–302.
42. Howe, *Reconcileableness*, pp. 2–4.
43. Howe, *Reconcileableness*, p. 52. Hammond's argument that prescience does not necessitate future actions is in *Pacifick Discourse*, pp. 2–4.
44. Hammond, *Pacifick Discourse*, pp. 39–41; Howe, *Reconcileableness*, pp. 100–105, 121–125.

Howe anticipated this question, and both argued that individuals *can* resist God's "super-effluent" grace, thus for all practical purposes affirming the Arminian position.[45]

There are two things about Howe's work I wish to emphasize. First, although early in his argument Howe warned against assuming the sufficiency of human reason to resolve all doctrinal difficulties, Howe's emphasis was on removing as many difficulties as possible by showing his doctrinal position to be a rational one. Indeed, Howe sounded more like Glanvill than he did his fellow nonconformists Baxter, Ferguson, and Owen when he argued that if we ascribe "inconsistencies" to or "give a self-repugnant notion" of God, we "gratifie atheistical minds." Even though Howe urged that when we encounter problems we should ask first whether the difficulty is in our own understanding and that we should be careful not to mistake a dictate of our depraved nature for a common notion, he also warned against using the limits of human understanding as an excuse not to attempt to resolve the apparent difficulty or inconsistency – and attempting to resolve the difficulty is precisely what Howe proceeded to do.[46] However, Howe failed to note (or was perhaps unaware of the fact) that Hammond, in 1660, had concluded his similar argument with the statement that Christians must assent to both God's prescience and man's free will, and that if someone were to judge Hammond's reconciliation of these two apparently contradictory truths inadequate, then that person should still assent to both truths in the "assurance that God can reconcile his own contradictions."[47]

Second, in *Reconcileableness*, Howe included a lengthy digression in which he objected to the Calvinist doctrine of absolute predestination on the grounds that such a doctrine makes God the author of sin.[48] Howe's claim was certainly not original. On many previous occasions, Calvinists had encountered similar arguments; as I noted earlier, in the

45. Hammond, *Pacifick Discourse*, p. 41; Howe, *Reconcileableness*, pp. 100–101.
46. Howe, *Reconcileableness*, pp. 6–7, 10.
47. Hammond, *Pacifick Discourse*, p. 160. Lorenzo Valla had come close to this position when he argued both for God's prescience and for man's free will, *and* noted that the two positions are contradictory; see his *Dialogue on Free Will* in *The Renaissance Philosophy of Man*, edited by Ernst Cassirer, Paul Oskar Kristeller, and John Herman Randall, Jr. (Chicago: University of Chicago Press, 1948). I discuss Valla's views in chapter 1. I thank James E. Force for bringing this work to my attention.
48. Howe, *Reconcileableness*, pp. 31–50.

Racovian Catechism, for example, the authors had argued that "since it is altogether necessary that sin should go before damnation, certainly he that decrees a man shall be of necessity damned, decrees also that he shal of necessity sin."[49] This digression did not escape the notice of Calvinists, who immediately responded.

The Ensuing Controversy

Immediately after the publication of Howe's *Reconcileableness,* the nonconformist Calvinist Theophilus Gale attacked Howe in Part 4 of his *Court of the Gentiles.* In *Reconcileableness,* Howe had argued that God concurs only in the good actions of individuals (and hence is not implicated in sinful actions).[50] In *Court of the Gentiles,* Gale responded by arguing that no action of any human being is unmixed with sin, and that because of this God cannot concur in the good part of an act without concurring in the sinful aspect at the same time.[51] Howe responded immediately with a *Postscript* in which he defended his position and reiterated that he had originally written about the subject at Boyle's request.[52] In the *Postscript,* Howe distinguished between "immediate" concurrence and "predeterminative" concurrence, stating that God immediately concurs in all actions but denying that God predeterminatively concurs in wicked actions.[53]

The following year, another nonconformist Calvinist, Thomas Danson, attacked Howe in the anonymous *De Causa Dei* (identifying himself only as "T.D."). In this work, Danson asserted that "God does determine, or move all Creatures to all and each of their actions," and argued that by making human beings the cause of some actions and displacing God as the first cause of all actions, Howe borders "as near upon *Arminianism* as *Scotland* does upon *England.*"[54] Andrew Marvell, the

49. *Racovian Catechism,* p. 146.
50. Howe, *Reconcileableness,* p. 44.
51. Theophilus Gale, *The Court of the Gentiles, Part IV, Of Reformed Philosophie, wherein Plato's Moral, and Metaphysic, or prime Philosophie is reduced to an useful Forme and Method* (London, 1677), p. 523.
52. "I wrote it upon the motion of that honourable Gentleman to whom it is inscribed; who apprehended somewhat of that kind might be of use to render our Religion less-exceptionable to some persons of an *enqiring* [sic] *disposition,* that might perhaps be too sceptical and pendulous, if not prejudic't" (Howe, *Reconcileableness, Postscript,* p. 2).
53. Howe, *Reconcileableness, Postscript,* pp. 28–29.
54. T[homas] D[anson], *De Causa Dei; or, A Vindication of the Common Doctrine of Protestant Divines, concerning Predetermination: (i.e., the In-*

poet, immediately (although anonymously) answered Danson and defended Howe. In *Remarks upon a Discourse*, Marvell noted that many divines "out of a vain affectation of Learning, have been tempted into Enquiries too curious after those things which the Wisdom of God hath left impervious to Humane Understanding."[55] Marvell's response was restrained and gentlemanly in tone; for example, he referred to Danson's work as "T.D." because he wanted to preserve the anonymity of the author of *De Causa Dei* and because he wanted to stress that he was attacking "the discourse" itself, and not its author.[56] Nevertheless, he went on to argue that no matter how much Calvinists might deny it, their doctrine of absolute predestination does indeed make God the author of sin.[57]

Howe's work had generated quite a controversy, and he had made it clear that he had written it at Boyle's request. It is not clear, in retrospect, just exactly what kind of work Boyle had anticipated Howe's producing. Given the similarity between Hammond's statement in *Pacifick Discourse* that "God can reconcile his own contradictions," and Boyle's position as it would be expressed in *Things above Reason,* it is possible that Boyle had intended for Howe to produce a work much like *Things above Reason.* It seems more likely, however, that Boyle had hoped that the time was right to present Hammond's attempted reconciliation of the Calvinist and Arminian positions once again, which is what Howe actually did. In any event, instead of producing a reconciliation, Howe's work ended up contributing to sectarian controversy, and Boyle went on, three years later, to publish his views on the limits of human reason. It is against this background that I turn now to Boyle's works on the limits of reason.

---

terest of God as the first Cause in all the Actions, as such, of all Rational Creatures:) from the invidious Consequences with which it is burdened by Mr. John Howe, in a late Letter and Postscript, of God's Prescience (London, 1678), pp. 3, 121.

55. [Andrew Marvell], *Remarks upon a Discourse writ by one T. D., under the pretence De Causa Dei, and of Answering Mr. John Howe's Letter and Postscript of God's Prescience, Affirming, as the Protestant's Doctrine, That God doth by Efficacious Influence universally move and determine Men to all their Actions, even to those that are most Wicked* (London, 1678), p. 3. On the title page, the author of this work is identified only as "a Protestant."

56. Marvell, *Remarks upon a Discourse*, pp. 14–15.

57. Marvell, *Remarks upon a Discourse*, p. 6.

4

Theology and the Limits of Reason

Boyle's *Discourse of Things above Reason* (1681) was not his first sustained treatment of the relationship between reason and religion, but it was his most complete and most sophisticated treatment. He devoted two earlier works to the topic, however, and it is appropriate to consider them briefly before moving on to *Things above Reason*. In addition, there were other works in which Boyle touched on the subject, and I incorporate passages from those works into the discussion in this chapter when doing so helps to flesh out Boyle's views. (I have already discussed similar passages from works prior to 1660 in the Introduction and in chapters 2 and 3.)

Style of the Scriptures

The first work that Boyle devoted to the relationship between reason and religion was *Style of the Scriptures* (1661). Originating from his unpublished manuscript "Essay of the holy Scriptures" (written c. 1651–1652 and discussed in chapter 2), *Style* was written to defend scripture from objections that it is obscure, that it is unmethodical, and that it is self-contradictory. In this defense, Boyle relied heavily on various considerations concerning the application of human reason to scripture, and because his intent was to foster the reading and study of scripture, his emphasis was for the most part on reason's competence to understand the content of scripture and on the importance of human learning to prepare reason to understand. There is a heavy emphasis on the importance of the study of languages and historical context, as well as on Boyle's claim that doctrines fundamental to salvation are "expressed with an evidence proportionable to the degree of assent that they exact, and are as far forth intelligible to pious and industrious readers, as they are necessary to be understood by them."[1]

1. Boyle, *Style*, *Works*, vol. 2, p. 167.

Despite the overall emphasis on reason's competence, however, Boyle also offered a number of reasons why some passages of scripture cannot be understood by some people, and why other passages simply cannot be understood at all. Because scripture was intended to serve the entire body of Christians past, present, and future, he claimed, it may well contain some passages intended for particular audiences. Some passages might have been most easily understood by people living in the distant past, while others (especially those pertaining to prophecies), might be understood by future generations. Further, some individuals might be intellectually capable of understanding difficult passages, whereas others need "plainer talk." Thus,

> God knowing that some persons must be wrought upon by reason, others allured by interest, some driven in by terrour, and others again brought in by imitation, hath by a rare and merciful (if I may so call it) suppleness of wisdom so varied the heavenly doctrine into ratiocinations, mysteries, promises, threats and examples, that there is not any sort of people, that in the scripture may not find religion represented in that form they are most disposed to receive impressions from.[2]

There are also passages, Boyle noted, in which the difficulty of understanding is not due to the style in which scripture was written, but "in the thing itself." After pointing out a number of questions of natural philosophy of a similar nature, such as the nature of space and time and the origin of motion, Boyle identified such theological questions as the nature and decrees of God, as well as such sublime mysteries as the Trinity and the Incarnation, as posing particular difficulties in understanding. Boyle believed that understanding might be enhanced by diligent study, but he also believed that understanding is often made more difficult by

> divers subtle men, who being persuaded, upon certain metaphysical notions they are fond of, or by the authority of such either churches or persons as they highly reverence, that such or such niceties are either requisite to the explication of this or that doctrine delivered in scripture, or, at least, deducible from it,

that they needlessly complicate the doctrines by twisting together "a revealed truth with their own metaphysical speculations about it." Although some parts of scripture are made obscure by this kind of nonsense, others are obscure in and of themselves, and understanding of some of the latter may be reserved to the "illumination and blazes of the last and universal fire." To find fault with the style in which scripture was written because of the obscurity of some passages was, he believed,

2. Boyle, *Style, Works,* vol. 2, pp. 262–263.

no less impertinent than to find fault with God for limiting human understanding. "It is," he concluded, "our duty to study" such passages of scripture, but it is not our duty to understand all of them.[3]

In Boyle's view, it is reason's role to discover how far its powers can reach and to discern those things to which reason's powers do not extend, and then to attempt to surmount difficulties that can be resolved, while wisely submitting itself to those that cannot be resolved.[4] And there is no doubt that in *Style*, his emphasis was on surmounting difficulties. Despite the passages concerning "difficulties" and "obscurities" in scripture, in *Style* Boyle expressed his most optimistic assessment of reason's competence in understanding revelation. *Style* was first conceived when he was in his early twenties, and was put into its final form when he was in his early thirties. It reflects the fruits of his study of Greek, Hebrew, Caldaic, Syriac, and Arabic, in which he had been encouraged by the Archbishop of Armagh, James Ussher.[5] Yet even in this early work there is considerable discussion of the limits of human understanding.

Reconcileableness of Reason and Religion

Style was directed toward Christians who acknowledged the authority of scripture but who shied away from studying it. In his next sustained treatment of the relationship between reason and religion, *Some Considerations about the Reconcileableness of Reason and Religion,* published in 1675, Boyle had a different audience in mind. In this work, he faced the same problem that had received St. Thomas's attention in the thirteenth century – the possibility that truths of natural philosophy might conflict with revealed truths. But whereas Aquinas had dealt with conflicts between the newly rediscovered Aristotelian philosophy and the Christian revelation, Boyle faced the charge that the newly rediscovered Epicurean mechanical philosophy might conflict with revealed truths. Despite the fact that Boyle believed that Pierre Gassendi had "baptized" Epicureanism, he was aware that "infidels" might deny God's existence and providence and support their views by appeal to Epicurean or other mechanical principles of philosophy. *Reconcileableness* was written to

3. Boyle, *Style, Works,* vol. 2, pp. 266–268; quotations from pp. 266 and 267.
4. Boyle Papers, vol. 1, fol. 64; the passage is quoted in full on page 100.
5. For Ussher's influence, see the excerpt from "Essay on the [holy] Scriptures" published in *Works,* vol. 1, p. xlviii-l, esp. p. xlviii. This excerpt from Boyle's unpublished manuscripts has since been lost. Also see Michael Hunter, "How Boyle Became a Scientist," *History of Science* 33 (1995), pp. 72–73.

convince such "infidels" (who argued that to be a Christian "one must cease to be a man, and much more, leave off being a philosopher") that the truly rational thing to do is to embrace the Christian revelation.[6]

Boyle's argument was far-ranging. In it, he limited religion to the truths found in scripture or legitimately deduced from scripture, specifically excluding other claims made by various sects of Christians – claims that might indeed involve false doctrines that are contrary to reason. Similarly, he limited philosophy to truths discoverable by natural reason per se, excluding "received opinions" of various sects of philosophers and noting that those opinions conflicted one with another. He also distinguished abstract reason from concrete reason, noting that the concrete reason exercised by individuals is subject to various impediments such as private interests and inadequacies of education; further, he appealed to impediments that are present from birth, appealing specifically to Francis Bacon's "idols of the tribe."[7] His primary point, however, was that the truths of natural philosophy must be limited to those cases from which and for which they were derived, a point we have seen made already in Stillingfleet's *Origines Sacrae* (1662) and in Ferguson's *Interest of Reason in Religion* (1675, the same year that Boyle published *Reconcileableness*). "There is a great difference," Boyle claimed,

> between a doctrine's being repugnant to the general and well-weighed rules or dictates of reason, in the forming of which rules, it may be supposed to have been duly considered; and its disagreeing with axioms, at the establishment whereof the doctrine in question was probably never thought on. . . . There may be rules, which will hold in all inferior beings for which they were made; and yet not reach to that infinite and most singular Being . . . and to some divine matters, which were not taken into consideration, when those rules were framed."[8]

Boyle was anxious to distinguish his argument from the double-truth argument (according to which a proposition might be acknowledged as

---

6. Boyle, *Reason and Religion, Works,* vol. 4, p. 151. For Gassendi's reception of Epicureanism, see Margaret J. Osler, "Baptizing Epicurean Atomism: Pierre Gassendi on the Immortality of the Soul," in *Religion, Science, and Worldview: Essays in Honor of Richard S. Westfall,* edited by Margaret J. Osler and Paul Lawrence Farber (Cambridge: Cambridge University Press, 1985), pp. 162–183; idem, *Divine Will and the Mechanical Philosophy: Gassendi and Descartes on Contingency and Necessity in the Created World* (Cambridge: Cambridge University Press, 1994), pp. 36–79. For the reception of atomism in England, see Robert Hugh Kargon, *Atomism in England from Hariot to Newton* (Oxford: Clarendon Press, 1966).
7. Boyle, *Reason and Religion, Works,* vol. 4, pp. 158–159, 164–171.
8. Boyle, *Reason and Religion, Works,* vol. 4, p. 161.

true in philosophy but false in theology, or vice versa), and he did so by reiterating that the truths of natural philosophy have been gathered from natural phenomena, and are therefore simply irrelevant where supernatural matters involving God's activity are concerned. Indeed, the truths of natural philosophy are rendered false whenever God intervenes in the usual course of nature. It was the failure to recognize this that resulted in the doctrine of double-truth, which was

> a great error, which has made so many learned men presume to say, that this or that thing is true in philosophy, but false in divinity, or on the contrary: as for instance, that a virgin, continuing such, may have a child, is looked upon as an article, which theology asserts to be true, and philosophy pronounces impossible.

This error, Boyle continued, was due to a misunderstanding. Natural philosophy does not declare the birth of a child to a virgin to be *absolutely* impossible, but simply impossible in the ordinary course of nature. When God's omnipotence is taken into consideration, however, it is not "repugnant to reason, that, if God please to interpose his power, he may."[9] When we consider God as the omnipotent author of the universe,

> we cannot but acknowledge, that, by with-holding his concourse, or changing these laws of motion, which depend perfectly upon his will, he may invalidate most, if not all the axioms and theorems of natural philosophy.[10]

Thus fire did not burn Daniel's three companions when they were cast into the fiery furnace; thus it is possible for God to resurrect the body even though it might be judged impossible according to the laws of nature. Assent to revealed truths would be irrational, Boyle argued, only if one were required to assent to something that it is impossible for God to do.

9. Boyle, *Reason and Religion, Works,* vol. 4, p. 163. The only sense in which Boyle acknowledged the occasional usefulness of the double-truth doctrine was in the sense that a philosopher might declare the proposition of natural philosophy to be definitely true, while admitting that a contradictory proposition of revelation *might* be true if that philosopher were to become convinced that it was a genuine revelation. Boyle's purpose here was to short-circuit the declaring of the revealed proposition to be false (ibid., p. 169). Presumably, a Christian philosopher would follow Boyle's admonition to consider the axiom of natural philosophy to have been superseded by God's intervention in nature. I discuss Boyle's view that an axiom of reason and a revelation contradictory to it might be both true in light of finite human understanding yet reconcilable in God's infinite understanding later in this chapter.

10. Boyle, *Reason and Religion, Works,* vol. 4, p. 161.

Generally, theologians have considered God's omnipotence to be limited only by the law of noncontradiction, and this was Boyle's view as well[11] – but, as the discussion of *Things above Reason* to follow will show, what he had in mind was not exactly what those other theologians had in mind.

Things above Reason

"One of the best, though least thought of Uses of Humane Reason, consists, *First,* in discovering how far its Powers can reach, and to what things they do not extend, and *then* in nobly attempting to surmount Difficulties that *are* superable, and wisely submitting itself to those that *are not* so."[12] In *Style of the Scriptures,* Boyle's emphasis was on the ability of human reason and learning to help the Christian better understand the content of scripture even as he acknowledged that revelations "difficult in themselves" transcend reason's limits. And in *Reason and Religion,* his emphasis was on the need to limit natural reason to natural phenomena, thereby protecting the mysteries of Christianity from being challenged by the principles of natural philosophy. His *Discourse of Things above Reason* and its accompanying *Advices in judging of Things said to Transcend Reason* (1681) are consistent with what he had written earlier, but in these later works, his emphasis was on those things to which the power of human reason does not reach.

I have already suggested – at least implicitly – one reason for Boyle's change of emphasis, and that was the controversy generated by John Howe's *Reconcileableness of God's Prescience of the Sins of Men, with the Wisdom and Sincerity of his Counsels, Exhortations, and Whatsoever Means He uses to prevent them* (1677). Boyle considered the apparently endless doctrinal quarrels among Christians to be a serious threat to the well-being of Christianity itself. It was all very well for divines such as Richard Baxter, John Owen, and Robert Ferguson to argue that some truths of Christianity transcend human understanding; as far as that was concerned, they spoke truth – and an important truth at that. It was a truth that protected the mysteries of Christianity from the rational scrutiny of atheists and infidels. However, it was not equally well for them to go on to argue about the correct interpretation of those very doctrines they themselves had argued reason cannot comprehend; by doing *this* they added weapons to the arsenals of atheists and infidels (not to mention the fact that they contradicted themselves).

11. Boyle, *Reason and Religion, Works,* vol. 4, p. 159.
12. Boyle Papers, vol. 1, fol. 64.

Further, we should not forget that the threat of Socinianism still loomed large. Although the Socinians acknowledged that some truths of Christianity are above reason, they denied that any such truths are contrary to reason. One of Boyle's goals in *Things above Reason* and *Advices* was to explore exactly what it means to say that no truth of Christianity is contrary to reason. That he had this aim will become clear when I discuss Boyle's category of "unsociable" truths.

The irenical intent of *Things above Reason* and *Advices* is illustrated by the fact that Boyle cast both in the form of a dialogue, and in a particular form of dialogue at that. None of Boyle's characters browbeats another, as was so often the case in the Socratic dialogues, and none is cast in the role of a "simpleton," as was the case in Galileo's *Dialogue Concerning the Two Chief World Systems* (1632). Instead, Boyle's characters are presented as treating one another as equals, and as working together to try to reach some consensus about the problem at hand. It is difficult to avoid concluding that Boyle intended this pattern of interaction to serve as a model for the discussion of disagreements among the Christians of his day.[13] As for the interlocutors themselves, Sophronius developed fully the argument that there are "things above reason." Eugenius was the one who was most willing first to ask for a more detailed explanation of a particular point, and then to agree with Sophronius. Pyrocles was the interlocutor most apt to raise objections (in the most respectful and gentlemanly of ways) and to resist being convinced. Timotheus vacillated between responding favorably to the argument that there are "things above reason" and acknowledging the validity of Pyrocles's objections. At the end of the dialogue, Timotheus and Eugenius had been persuaded fully that there are "things above reason," and Pyrocles was inclined to admit that there may be such a category of propositions. In *Advices*, Arnobius took the place of Sophronius; the other three interlocutors remained the same.

Boyle's expressed purpose in *Things above Reason* was to assert that where both theology and natural philosophy are concerned, the proper role of human reason is to ascertain and acknowledge its limits. Boyle, aware that he could be accused of setting reason "awork to degrade itself," argued that he was instead exercising its "most noble and genuine

13. For an astute discussion of Boyle's use of the dialogue form as a means of persuasion, see Steven Shapin and Simon Schaffer, *Leviathan and the Air Pump: Hobbes, Boyle, and the Experimental Life* (Princeton: Princeton University Press, 1985), pp. 74–75. As Shapin and Schaffer note (p. 75), Boyle explicitly stated in *Sceptical Chymist,* also written in dialogue form, that he was "not sorry to have this opportunity of giving an example, how to manage even disputes with civility" (*Works*, vol. 1, p. 462).

exercise." He was showing that it is but a "limited faculty . . . [and that it is] an injury to the author of it, to think man's understanding infinite like his."[14]

I reserve discussion of Boyle's views on the limits of human understanding in the context of natural philosophy for Part II; in this section, I confine myself to Boyle's views on reason's limits in the context of theology, drawing not only on *Things above Reason* and *Advices*, but also on *Reflections upon a Theological Distinction,* an essay that accompanied Boyle's *Christian Virtuoso* (1690) and in which he iterated his views on things above reason, as well as other essays in which he addressed the issue.

Three Categories

In theology, things that are above reason are a subcategory of "such notions and propositions, as mere reason – that is, reason unassisted by supernatural revelation – would never have discovered to us."[15] Even after these notions and propositions are revealed, we cannot understand fully some of them – those that are above reason – for one of three reasons: We cannot form adequate ideas of the thing revealed (the truths are "incomprehensible"), or we cannot explain how certain effects are caused (the truths are "inexplicable"), or one truth contradicts something else that we know to be true or is self-contradictory ("unsociable" truths).[16] These categories are not mutually exclusive; any given truth may fall into two or even all three categories.[17] As a group, the truths in all three categories are called "priviledged things."[18]

God's nature is an example of a truth in the first category; it is incomprehensible to us.[19] How God made the world out of nothing, or how he unites an immaterial soul to a human body and maintains that union are among those things that Boyle termed "inexplicable."[20] Boyle's most common example of two "unsociable" (or contradictory) truths was that both the proposition that God has foreknowledge of the future and the proposition that man has free will are true; in addition, he argued that various attributes of God may appear contradictory to

14. Boyle, *Things above Reason, Works,* vol. 4, p. 411.
15. Boyle, *Reflections upon a Theological Distinction, Works,* vol. 5, p. 542.
16. Boyle, *Things above Reason, Works,* vol. 4, p. 407.
17. Boyle, *Things above Reason, Works,* vol. 4, p. 423.
18. Boyle, *Things above Reason, Works,* vol. 4, p. 409.
19. Boyle, *Things above Reason, Works,* vol. 4, p. 409.
20. Boyle, *Things above Reason, Works,* vol. 4, p. 423.

human beings.[21] A truth that falls into the category of "unsociable" truths may also be said to be a thing "contrary to reason."

Boyle went on to explain just how it is that these "privileged things" (and, most particularly, the category of "unsociable" truths) are exempt from "the rules and axioms and notions, whereby we judge of the truth and falshood of ordinary, or other things."[22] He did this by dividing the rules of reasoning into two categories. In the first category are the axioms that are applied to particular fields; these axioms are "physical propositions" that are but "limited, collected truths, gathered from the phenomena of nature." Boyle termed these "gradual" or "emergent" truths, and offered, as an example, the principle *ex nihilo nihil fit*.[23] Privileged things, Boyle claimed, are "of so heteroclite a nature [that they] may challenge an exemption from some of the rules employed about common things."[24] These "gradual" or "emergent" truths that are abstracted from the phenomena of nature should not be applied to supernatural things (revelation). (This point, of course, is the one that Boyle had made in *Reconcileableness of Reason and Religion*, that Ferguson had made in his *Interest of Reason and Religion*, and that Stillingfleet had made in his *Origines Sacræ*.)

The second category of rules of reasoning are those that spring from "the rational faculty itself, furnished with the light, that accompanies it, when it is rightly disposed and informed," such as that snow is white, that from truth nothing but truth can be deduced, and that contradictory propositions cannot both be true.[25] The light that accompanies this rational faculty "enable[s] a man, that will make a free and industrious use of it . . . to pass a right judgment of the extent of those very dictates that are commonly taken for rules of reason."[26]

Law of Noncontradiction

In considering the objection that nothing is more unreasonable than to embrace opinions that violate the very rules of reasoning, Boyle argued

21. Boyle, *Things above Reason, Works,* vol. 4, pp. 409, 424; *Advices, Works,* pp. 450, 465; *Reflections upon a Theological Distinction, Works,* vol. 5, p. 543. Boyle also stressed that God's nature is incomprehensible to human beings in a number of other works; see, for example, *Final Causes, Works,* vol. 5, p. 439 and *Reason and Religion, Works,* vol. 4, p. 161.
22. Boyle, *Things above Reason, Works,* vol. 5, p. 408.
23. Boyle, *Advices, Works,* vol. 4, pp. 462–463.
24. Boyle, *Advices, Works,* vol. 4, p. 451.
25. Boyle, *Advices, Works,* vol. 4, p. 458.
26. Boyle, *Advices, Works,* vol. 4, pp. 461–462.

that even the principles of logic may be considered to be but "gradual" or "emergent" truths from God's point of view because our "inbred or easily acquired ideas and primitive axioms ... do not extend to all knowable objects whatsoever, but reach only to such ... as God thought fit to allow our minds in their present (and perchance lapsed) condition."[27] Boyle would not commit himself as to the exact nature of reason's inability to comprehend fully all truths, but he most often appealed to a comparison of man's finite wisdom with God's infinite wisdom to support his argument.[28] He did not, however, deny that there might be other factors as well, such as Adam's fall or the way in which God has united individual souls with individual bodies.[29] In any event, God has helped human beings to ascertain the limits which he has set on human understanding by furnishing us with "a domestick monitor, or a kind of internal criterium" that allows us to discern that some matters are quite beyond our reach.[30]

To argue that the content of revelation is above reason and something that human beings cannot understand fully was not at all a new or unusual position for Boyle to take. But to argue that the content of revelation can violate the law of noncontradiction – can actually be *contrary* to reason – is a different matter. That any conclusion whatsoever follows from contradictory premises was well-known in Boyle's day,[31]

27. Boyle, *Things above Reason, Works*, vol. 4, p. 445.
28. Boyle, *Things above Reason, Works*, vol. 4, p. 410; *Reflections upon a Theological Distinction, Works*, vol. 5, p. 542. Similar passages may be found in *Excellency of Theology, Works*, vol. 4, p.21; *Christian Virtuoso, Works*, vol. 5, p. 515; *Reason and Religion, Works*, vol. 4, pp. 159, 161; *High Veneration, Works*, vol. 5, p. 153. I discuss the relationship between man's finite reason and Boyle's voluntarism in chapter 8.
29. Boyle, *Things above Reason, Works*, vol. 4, p. 445.
30. Boyle, *Things above Reason, Works*, vol. 4, p. 446. In *Things above Reason*, Boyle repeated his claim that God has provided human beings with the means of discerning reason's limits a number of times. The means is described as "a distinct and unwonted kind of internal sensation" (p. 417). Boyle claimed that "the great and free author of human nature, God, so framed the nature of man, as to have furnished his intellective faculty with a light, whereby it can ... discern some at least of the limits, beyond which it cannot safely exercise its act of particularly and peremptorily judging and defining" (p. 418). This is directly opposed to John Owen's claim that due to its depravity, reason is unable to discern its own depravity; see my discussions of Owen in chapter 2.
31. William Kneale and Martha Kneale, *The Development of Logic* (New York: Oxford University Press, 1962), p. 314.

and Boyle himself noted the difficulty involved when he observed that the rational soul of a Christian

> cannot avoid admitting some consequences as true and good, which she is not able to reconcile to some other manifest truth or acknowledged proposition. And whereas other truths are so harmonious, that there is no disagreement between any two of them, the heteroclite truths appear not symmetrical with the rest of the body of truths, and we see not how we can at once embrace these and the rest, without admitting that grand absurdity, which subverts the very foundation of our reasonings, that contradictories may both be true.[32]

Boyle was not at all ambiguous about the claim that two inconsistent propositions might both be true. In response to the charge (from Pyrocles) that nothing is "more clear to human understanding, or more supposed in almost all our ratiocinations, than that two truths cannot be contradictory to each other," Boyle responded by arguing that "either the propositions said to be repugnant are both really true, or they are not." If they are not, then there is no difficulty. But if both *are* true, then they must be reconcilable. And since it is assumed in this case that human reason cannot reconcile them, then it must be the case that "a superior intellect, such as unquestionably the divine is, can discover an agreement, wherein we cannot discern it."[33] In short, "where priviledged things are concerned, we are not always bound to reject every thing, as false, that we know not how to reconcile with some thing, that is true."[34]

Boyle scholars have generally failed to perceive this extreme expression of reason's limits. M.S. Fisher, for example, has argued that Boyle's position is that reason "is to give way only to admit revelation, upon revelation's entry reason is to return. In case of conflict between the two, if the question involves the fundamentals of reason, then the latter is to prevail."[35] In making this claim, Fisher relied on a passage in the *Appendix* to *The Christian Virtuoso*. In this passage, one of the interlocutors argued that if a revelation is interpreted in such a way as to contradict a fundamental axiom of reason, then the relevant passages of scripture must be reinterpreted in such a way as to *not* contradict the axiom, for

32. Boyle, *Things above Reason, Works*, vol. 4, p. 423.

33. Pyrocles's objection, *Advices, Works*, vol. 4, p. 462; Boyle's response, ibid., pp. 465–466.

34. Boyle, *Advices, Works*, vol. 4, p. 464 (emphasis deleted).

35. M.S. Fisher, *Robert Boyle, Devout Naturalist: A Study in Science and Religion in the Seventeenth Century* (Philadelphia: Oshiver Studio Press, 1945), p. 126.

"God, being infinitely knowing, and being the author of our reason, cannot be supposed to oblige us to believe contradictions."[36] It is important to note, however, that the *Appendix* was not prepared for publication by Boyle himself; it was assembled from fragments of Boyle's unpublished writings prior to the publication of the 1744 edition of the *Works*. In an unpublished "Introduction to my loose Notes Theologicall," Boyle noted that he jotted down many thoughts that he considered would be helpful later in formulating the arguments of various speakers in essays that would take the form of dialogues, "which Declaration will, I presume, keep you, as I desire it should," he continued, "from thinking that every passage thus introduc'd contains nothing but my own Opinion."[37] Indeed, the passage in the *Appendix* can be found in the Boyle Papers, in an area in which surrounding folios have the interlocutors' names penciled in.[38] Clearly, passages from the *Appendix* dealing with reason's limits or competence should be cited only when they agree with Boyle's views as expressed elsewhere, especially in light of the fact that he published a sustained discussion of his views on reason's limits only a year before his death.[39] Similarly, L.T. More missed the force of Boyle's views concerning reason's limits because he relied heavily on *Reasons Why a Protestant Should not Turn Papist* (1687), an essay long attributed to Boyle, but that has recently been shown to be the work of someone else.[40]

When Boyle claimed that although there are revelations that are above reason but not contrary to reason, he meant that they are not contrary to divine wisdom. In *ordinary* cases, in which neither of two contrary propositions is revealed, the law of noncontradiction holds for human understanding; in such cases, it is "the usefullest *criterion* to discriminate

36. Boyle, *Appendix* to *Christian Virtuoso*, *Works*, vol. 6, p. 712.
37. Boyle Papers, vol. 5, fol. 27; the title of the "introduction" appears on fol. 25.
38. Boyle Papers, vol. 1, fol. 82. Similar passages occur on fols. 165 and 166.
39. Boyle published *Reflections upon a Theological Distinction* in 1690; he died in 1691.
40. L.T. More, *The Life and Works of the Honourable Robert Boyle* (New York: Oxford University Press, 1944), esp. pp. 168 and 185. Edward B. Davis has recently identified the author of *Reasons why a Protestant should not turn Papist* as a former Jesuit turned Protestant; see "The Anonymous Works of Robert Boyle and the *Reasons Why a Protestant Should not Turn Papist* (1687)," *Journal of the History of Ideas* 55 (1994), esp. pp. 622–629. I discuss other misinterpretations of Boyle's views on reason's limits and the consequences of those misinterpretations in the Conclusion.

between falshood and truth."[41] But if one of the propositions is a revealed one (and has been proven by argument sufficient in its kind to have been revealed), and if human reason cannot discern the harmony between the truths, then both should be accepted as true in the assumption that God perceives the harmony among all truths. And this, Boyle claimed, is exactly the case in the clash between the truths that God has foreknowledge of future events and that human beings have free will.

Prescience and Free Will

In his numerous discussions of prescience and free will, Boyle was unambiguous in claiming both that the two relevant propositions are indeed, from the human perspective, contradictory, and that both are true. That God has foreknowledge, he argued, has been revealed in scripture by fulfilled prophecies. Further, it is rationally demonstrable on the grounds of God's perfections. That human beings possess free will, he claimed, is an obvious truth because "they have felt it in themselves."[42] Yet the "greatest wits" that have attempted to reconcile these two truths have been unable to do so without maintaining "something or other, that thwarts some acknowledged truth or dictate of reason." These two propositions, Boyle continued, "afford us an instance of truths, whose consistency and whose symmetry with the body of other truths our reason cannot discern, and which therefore ought to be referred to that sort of things above reason, that I call unsociable."[43] And in *Reflections upon a Theological Distinction* (1690), Boyle again made the point that although it is true that God has prescience and that human beings have free will, human understanding is not able to reconcile these two "unsociable" truths.[44] In short, according to Boyle, privileged truths sometimes involve things that are "above logic."[45]

Strictly speaking, Boyle avoided violating the law of noncontradiction by arguing that God, in his infinite wisdom, perceives the harmony among all truths. And yet Boyle did "subvert the very foundation of our reasonings" when he asserted that two propositions that appear to human beings to be truly contradictory may both be true. If we look only at Boyle the natural philosopher, we might think that in going so

41. Boyle, *Advices, Works,* vol. 4, p. 466.
42. Boyle, *Advices, Works,* vol. 4, p. 466.
43. Boyle, *Things above Reason, Works,* vol. 4, p. 424.
44. Boyle, *Reflections upon a Theological Distinction, Works,* vol. 5, p. 543.
45. Boyle, *Advices, Works,* vol. 4, p. 448.

far he was simply trying to protect the mysteries of religion from being challenged by any findings of the new natural philosophy. Indeed this was probably what he had in mind when he said that a revealed truth might *appear* contradictory to those "emergent" or "gradual" truths that are gathered from the phenomena of nature. As we have seen, however, finite human reason is able to eliminate this apparent contradiction by limiting truths gathered from natural phenomena to the objects of natural knowledge and recognizing that they may be superseded by supernatural phenomena. But when he went on to argue that some revelations (as understood by finite beings) actually *are* exempt even from the law of noncontradiction, he had in mind the theological controversies of his day. Boyle's *Things above Reason* was his answer both to the claim of the Socinians that no article of the Christian faith should be interpreted in such a way as to be "contrary to reason" and to the claim of the Calvinists that any emphasis on man's free will and man's role in the salvation process violates what is known about both God's prescience and his power.

The Charge of Enthusiasm and *Advices*

Boyle was aware that in seventeenth-century England, anyone who was perceived as denigrating human reason was vulnerable to charges of enthusiasm. We have already seen that when Ferguson argued reason's limits, Glanvill accused him of being an enthusiast; might not Boyle be similarly accused? Boyle, after all, had gone even further than Ferguson in arguing reason's limits; Boyle had gone so far as to argue that from the point of view of finite human understanding, two contradictory propositions might both be true, an argument he himself realized could "subvert the very foundation of our reasoning." And in fact, Henry Hallywell, writing just two years after the publication of Boyle's *Things above Reason* and *Advices*, complained of those who "make the choicest of [religion's] Articles so incomprehensible as to be elevated above Reason," and argued that nothing can "give greater Ground to the bold Cavils and Pretentions of Enthusiasts and disguised Atheists."[46]

Boyle anticipated that his argument in *Things above Reason* could be construed as providing grounds for the claims of enthusiasts, or, as he put it, for "rash or imposing men" to forward any personal beliefs or

---

46. Henry Hallywell, in his annotations to George Rust's *A Discourse of the Use of Reason in Matters of Religion: Shewing, That Christianity Contains*

any interpretations of scripture whatsoever, and to defend such beliefs or interpretations by simply claiming that they were both revealed and "above reason" and hence could be neither rationally defended nor attacked (which, in fact, was exactly the way both Robert Ferguson and John Owen had defended their own Calvinist interpretations of scripture in the 1670s).[47] Boyle's *Advices in judging of Things said to Transcend Reason* was intended to anticipate such objections.

In *Advices,* he noted that he could not insist that many revealed matters are "above reason" and that some are even "contrary to reason" and at the same time totally exclude the possibility of being "imposed on by ourselves or others, when such sublime subjects are treated or discoursed of."[48] For Boyle, the most important question was not whether his argument "may possibly be misapplied by rash or imposing men, but whether it be grounded on the nature of things."[49] Boyle knew that his *Advices* could, at best, minimize the danger of such an imposition. But he also knew that such impositions could always be countered by a *tu quoque* argument. After observing that there will always be

> men, who either out of ignorance and passive delusion, or out of self-confidence, or out of design, take upon them, with great bold-ness, to affirm what they please about priviledged subjects; and when they are opposed in their extravagancies by ratiocinations

---

nothing *Repugnant to Right Reason; Against Enthusiasts and Deists* (London, 1683), p. 49. Hallywell admitted that there are "things above reason" only in the sense that are some propositions that human reason unaided by revelation could not have discerned. Once revealed, however, the propositions are not incomprehensible (pp. 49–50). An individual who believes what cannot be understood, Hallywell argued, "believes he knows not what. ... While he dreams of believing some unintelligible Mystery, he only pursues mere shadows of words" (pp. 26–27). Although Hallywell included a brief passage in which he acknowledged that some revelations may exceed a particular individual's current level of understanding (p. 41), this does not temper the tone of his argument, which shows that he believed it foolish to assent to that which one cannot comprehend. On enthusiasm, see Michael Heyd, "The Reaction to Enthusiasm in the Seventeenth Century: Towards an Integrative Approach," *Journal of Modern History* 53 (1981), pp. 258–280.

47. For documentation of "rash and imposing," see note 49. In *Reflections,* Boyle himself had noted that some individuals had in fact used the argument that certain propositions are above reason "to elude some objections, that cannot otherwise be answered" (pp. 541–542).
48. Boyle, *Advices, Works,* vol. 4, p. 448.
49. Boyle, *Reflections upon a Theological Distinction, Works,* vol. 5, p. 547.

they cannot answer, they urge, that these things being above reason,
are not to be judged of by it.

The correct answer to such claims, Boyle argued, is to respond that if
the belief being proposed is truly above reason, then "it will surpass [the
proposer's reason] as well as mine, and so leave us upon even terms."[50]

His intention was not, he noted, to be "so vain as to pretend to frame
a logic about things above logic." Instead, he would offer "not rules, but
advices . . . [not] so much as directions to find the truth, in such abstruse
matters, as cautions that may assist" individuals in avoiding "some
errors and mistakes."[51] The first "advice" was "that about privileged
subjects themselves, we do not admit any (affirmative) assertion, without
such proofs, to evince it, as are sufficient in their kind," and he offered
God's prescience as an example of a sufficiently evidenced assertion.[52] In
this first "advice," he argued that although the revelation itself may be
beyond reason's sphere, if individuals are to hold any doctrinal beliefs at
all, they often must accept one of two or more conflicting interpretations
of scriptural revelations – and it is reason's role to evaluate the arguments offered for the various interpretations:

> About a priviledged thing, as well as about any other, propositions
> may be framed, and often are so, that are contrary to one another;
> [and] to assent to both were to be sure to believe one falsity, if not
> two. And if we will assent to but one, we must either judge at
> adventures, or allow ourselves to examine the mediums of proba-
> tion, employed on both sides, and thereupon judge, why one of the
> propositions is to be assented to and the other rejected.[53]

In the *Appendix* to the *Christian Virtuoso* (1690), Boyle offered a criterion for choosing between alternative interpretations. He did not, he
said, consider himself obliged

> to have the same regard and respect for the explications, that the
> schoolmen and many other divines give us of the mysteries of Chris-
> tianity, that I have for the articles themselves; since for the mysteries
> I have the divine authority of the revealer, who can oblige my
> faith to assent even to dark truths: but for the expositions and
> consequences, I have but human authority: and though clearness is

---

50. Boyle, *Advices*, *Works*, vol. 4, p. 449.
51. Boyle, *Advices*, *Works*, vol. 4, p. 448.
52. Boyle, *Advices*, *Works*, vol. 4, pp. 449–450. Elsewhere, Boyle argued that
     God's prescience is rationally deducible from scripture, basing the argu-
     ment on scriptural prophecies and on God's infinite perfections (*Reason
     and Religion*, *Works*, vol. 4, p. 176; see also *High Veneration*, *Works*, vol.
     5, p. 151).
53. Boyle, *Advices*, *Works*, vol. 4, p. 449.

not always necessary to divine mysteries, yet it may be justly exacted
in a good explication, since to explicate a thing, is to render it at
least intelligible.[54]

Even so, in assenting to one proposition and rejecting the other, an
individual should be very aware that the arguments involved do not
amount to rigid demonstrations.[55] An individual should assent to a
particular interpretation of a scriptural passage only provisionally, con-
stantly bearing in mind that the limits of human understanding prohibit
that individual's being absolutely sure that the interpretation assented to
is the one uniquely correct one.

Boyle's second "advice" was "that we be not hasty to frame negatives
about priviledged things, or to reject propositions or explications concerning them; at least, as if they were absurd or impossible." And if
questioned whether he has, in his second "advice," contradicted what he
argued in his first "advice," he would answer that

> there is a great deal of difference between believing a proofless
> affirmation about things, which the affirmer does not know to be
> true, and framing negative conclusions against opinions, which, for
> aught we yet clearly know, may be true.

Where the second "advice" is concerned, Boyle was careful to note an
exception. In a passage that seems to be clearly directed against Hobbes
(and perhaps Spinoza as well), Boyle argued that if there is a clear
revelation concerning one of God's attributes, the negative of that attribute should be denied; hence, although human understanding cannot
understand fully God's nature, human understanding can discern that
God is immaterial, and therefore cannot be of a corporeal nature.[56]
Despite this exception, his primary point in the second advice was that
an individual should not be overly hasty in rejecting another individual's
interpretation of scriptural revelation as false. As he put it,

> it will become us at least to forbear a rude insulting way of rejecting
> the opinion of learned men, that dissent from us about such things;
> since the sublimity of the subject should make mistakes about them

---

54. Boyle, *Appendix to Christian Virtuoso, Works,* vol. 6, p. 713.

55. Boyle, *Advices, Works,* vol. 4, p. 450.

56. Boyle, *Advices, Works,* vol. 4, p. 453. For contemporary conflation of
    Hobbes's materialism and Spinoza's pantheism, see Michael Hunter, *Sci-
    ence and Society in Restoration England* (Cambridge: Cambridge Univer-
    sity Press, 1981; Aldershot, Hampshire: Gregg Revivals, 1992), pp.
    169–170. See also Rosalie L. Colie, "Spinoza in England 1665–1730,"
    Proceedings of the American Philosophical Society 107 (1963), pp. 183–
    219.

the more easy to be pardoned, because they are difficult to be avoided.[57]

If one is tempted to "rudely" or "insultingly" reject a fellow Christian's interpretation of difficult scriptural passages, Boyle recommended instead either that judgment be suspended or that "a wary and unprejudiced assent [be given] to opinions, that are but faintly probable."[58]

Boyle's third "advice" was a restatement of his claim in *Things above Reason* that where privileged things are concerned, individuals ought not deny that something is true simply "because we cannot explain, or perhaps so much as conceive, the modus of it."[59] After giving several examples from "things corporeal," Boyle offered "the divine prescience of future contingents" as a theological example. Although it would be "impious to deny" that God has foreknowledge of future contingents, finite human beings do not understand how he has such knowledge. Another theological example of an "inexplicable" revealed proposition was "how in [God] there can be a Trinity."[60] This "advice" seems to be directed primarily against the Socinians, who, as I noted in chapter 3, denied both the orthodox interpretation of the Trinity and God's foreknowledge of future contingents.[61]

In his fourth, fifth, and sixth "advices," Boyle dealt with the category of things "above reason" that from the point of view of finite human understanding are either self-contradictory or contradictory to some other known truth. The fourth "advice" was "that when we treat of priviledged subjects, we are not bound always to think every thing false, that seems to thwart some received dictate of reason."[62] In his discussion of this "advice," Boyle expounded on the argument he had offered in *Things above Reason* that some revelations are exempt not only from the axioms of this or that particular science, but are even exempt from the rules of reasoning themselves. Concerning these revelations, Boyle claimed that

> the native light of the mind may enable a man, that will make a free
> and industrious use of it, both to pass a right judgment of the extent

---

57. Boyle, *Advices, Works*, vol. 4, p. 452.
58. Boyle, *Advices, Works*, vol. 4, p. 452.
59. Boyle, *Advices, Works*, vol. 4, p. 453.
60. Boyle, *Advices, Works*, vol. 4, pp. 454, 455.
61. Boyle was well aware that the Socinians denied both of these propositions. For his awareness of the Socinian denial of the Trinity, see my discussion of his "Essay of the holy Scriptures" in chapter 2. In *Reason and Religion*, Boyle noted that the Socinians deny God's prescience (*Works*, vol. 4, p. 176).
62. Boyle, *Advices, Works*, vol. 4, p. 456.

of those very dictates, that are commonly taken for rules of reason,
and to frame others on purpose for priviledged things.[63]

Boyle's primary theological example here was that where everything
except God is concerned, existence is separable from essence; in God,
however, the two are inseparable.[64]

Boyle's fifth "advice" was "that where priviledged things are concerned, we are not always bound to reject every thing, as false, that we
know not how to reconcile with some thing, that is true."[65] Here Boyle's
theological example was God's prescience and man's free will,[66] so this
"advice" would seem to be aimed directly at anyone (such as Hobbes,
Spinoza, or staunch Calvinists) who denied man's free will, as well as at
anyone (such as the Socinians) who denied God's prescience.

Boyle's sixth "advice" was "that in priviledged things we ought not
always to condemn that opinion, which is liable to ill consequences, and
incumbred with great inconveniencies, provided the positive proofs of it
be sufficient in their kind."[67] Boyle's point here was that some individuals reject a proposed explication of a scriptural passage on the grounds
that it leads to "ill consequences." Nevertheless, any alternative explication might also lead to insurmountable difficulties. "We must not,"
Boyle argued, "expect to be able, as to priviledged things, and the
propositions that may be framed about them, to resolve all difficulties,
and answer all objections."[68] With this "advice," Boyle most likely had
in mind objections such as those of the Calvinists that an emphasis
on free will is fraught with the "ill consequence" of limiting God's
omnipotence, or those of Anglicans that an emphasis on absolute and
irrespective decrees leads to antinomianism.

Conclusions

Before moving on to Part II and Boyle's views on the limits of human
reason in the context of natural philosophy, I will pause to consider
some of the more important points that can be gleaned from Boyle's
writings concerning things above reason in the context of theology. It is
important to note that in his writings about things above reason Boyle
drew not only on his own early (and to a great extent unformed) views

63. Boyle, *Advices, Works,* vol. 4, pp. 461–462.
64. Boyle, *Advices, Works,* vol. 4, p. 463.
65. Boyle, *Advices, Works,* vol. 4, p. 462.
66. Boyle, *Advices, Works,* vol. 4, pp. 455–456.
67. Boyle, *Advices, Works,* vol. 4, p. 467.
68. Boyle, *Advices, Works,* vol. 4, p. 469.

of the limits of reason as expressed in his unpublished "Essay on the holy Scriptures," but also on the writings of those individuals who had engaged in the controversy over the limits of reason during the 1650s, 1660s, and 1670s. In arguing that human reason is not competent to judge the content of God's revelation, Boyle aligned himself with what was predominantly (though not exclusively) a nonconformist position. Boyle's category of "inexplicable" things above reason echoed Richard Baxter's argument that human beings cannot understand "how these things can be," as well as Boyle's own thoughts about the inability of human beings to understand the "modus" or "manner" of things as expressed in the early "Essay." In his discussion of "incomprehensible" things, Boyle expanded his own thoughts as expressed in "Essay" in such a way as to echo Robert Ferguson's discussion about the error of assuming that the content of revelation includes only ideas that can be "clearly and distinctly" perceived.

Boyle, in his discussion of "emergent" or "gradual" truths, further echoed Ferguson, and extended Ferguson's comments to include even the principles of logic, which, Boyle argued, might be considered "emergent" or "gradual" truths from God's perspective. In doing this, Boyle provided a rationale for the category of "unsociable" things, a category that had only been hinted at earlier when Baxter had argued that human beings cannot perceive the harmony among all truths, when Hammond had stated that God can reconcile his own contradictions, and when Owen had stated that some revelations may indeed be contrary to reason insofar as reason is debased. I am not claiming that Boyle had read all of the particular works to which I have referred; nevertheless, either he had read them or had encountered similar arguments elsewhere. At the beginning of *Reflections upon a Theological Distinction,* a work in which Boyle restated the views he had expressed earlier in *Things above Reason,* he noted (concerning the distinction that some things are said to be "above reason" and others are said to be "contrary to reason") that

> as far as I can discern by the authors, wherein I have met with it, (for I pretend not to judge of any others) there are divers that employ this distinction, [but] few who have attempted to explain it, (and that I fear not sufficiently,) and none that has taken care to justify it.[69]

It should be noted that the tracing of specific influences is not the point here. It is possible that both Boyle and the nonconformists who

69. Boyle, *Reflections upon a Theological Distinction, Works,* vol. 5, p. 542. John Harwood, editor of *The Early Essays and Ethics of Robert Boyle.*

wrote urging reason's limits shared similar sources. What is important is that he did not hesitate to align himself with what was at the time basically a nonconformist position, even at the risk of leaving himself open to charges of enthusiasm. This fact may modify our views of Boyle as an "orthodox" Anglican and as a latitudinarian, depending, of course, on how we define "orthodoxy" and "latitudinarianism."

It is also important to note that even while aligning himself with the arguments of the nonconformists concerning reason's limits, Boyle did not align himself with the implications of those arguments urged by some nonconformists. While sharing their desire to take the wind out of the Socinians' sails, Boyle did not, as did Owen and Ferguson, use arguments concerning reason's limits to shield Calvinist interpretations of disputed scriptural passages from the rational scrutiny of Anglican opponents. In fact, Boyle used *Things above Reason* and *Advices* to express his own belief that the Anglican interpretation of predestination as conditional was probably the correct interpretation, although he did so in such an understated and nondogmatic way as to avoid controversy. The one theological example that permeates these works from beginning to end is that of God's prescience and man's free will to illustrate "unsociable" truths. Had he so desired, Boyle could have used God's absolute and unconditional decrees and man's free will as his example. By stressing prescience and not mentioning irrespective decrees, Boyle was (at least provisionally) rejecting the Calvinist interpretation of predestination and aligning himself with Anglicans and the more moderate nonconformists. Boyle's "Love of peace" prohibited his participation in this (or any other) specific theological controversy; nevertheless, his "love of truth" caused him to affirm his own belief in conditional predestination in his choice of examples.

I term Boyle's rejection of the Calvinist view "provisional" because doing so points to Boyle's primary intention in *Things above Reason* and *Advices*. As Benjamin Calamy, a nonconformist himself, remarked

---

(Carbondale and Edwardsville: Southern Illinois University Press, 1991), has published what he claims is a partial listing of the works in Boyle's library (pp. 249–281). I do not believe that the evidence warrants Harwood's claim that Royal Society MS. 23, which is the basis of Harwood's claim, is a list of works owned by Boyle. Instead, it is far more likely that it is a list of works owned by John Warr, Jr., Boyle's longtime servant and one of the executors of his will. Michael Hunter argues convincingly that the library is probably Warr's in *Letters and Papers of Robert Boyle: A Guide to the Manuscripts and Microfilm* (Bethesda, MD: University Publications of America, 1992), pp. xxi–xxii.

in 1673, "it were much to be wished that men had never gone about to explain those things which were mysterious, and to give an account of those things which [they] themselves acknowledg'd incomprehensible."[70] This was Boyle's wish as well. Although Boyle acknowledged that it is possible and permissible to discourse of "priviledged things" rationally, it was, he claimed, "arrogance" for an individual to speak of such matters with the same confidence one might have in speaking of other matters. Things above reason should be discussed only with "a peculiar wariness, and modest diffidence."[71]

Boyle made it clear that where something "above reason" is concerned, there *is* a uniquely correct understanding; his point was that human beings can never be certain they have attained that understanding. The fact that a particular matter is above reason is "extrinsical and accidental to its being true or false," Boyle pointed out in *Reflections upon a Theological Distinction,* and went on to explain that

> to be above our reason is not an absolute thing, but a respective one, importing a relation to the measure of knowledge, that belongs to the human understanding, such as it is said to transcend: and therefore it may not be above reason, in reference to a more enlightened intellect, such as in probability may be found in rational beings of an higher order, such as the angels, and, without peradventure, is to be found in God.[72]

Angels may know whether the opinion being urged is true or false, and God most certainly knows. Individual human beings, however, can only seek the truth and judge alternative claims concerning truth as best they can. Although God need not pardon those who fail to seek the truth, Boyle believed that God does pardon "those seekers of his truths, that miss them."[73] Meanwhile, individuals should not be hasty to reject the opinions of those who disagree with them about such matters because "the sublimity of the subjects should make mistakes about them the more easy to be pardoned," and the difficulty in penetrating such abstruse matters should keep us from being "over-confident, that we also may not be mistaken."[74]

70. Benjamin Calamy, *A Sermon Preached before the Right Honorable the Lord Mayor and the Court of Aldermen at Guild-Hall Chappel upon the 13th of July, 1673* (London, 1673), p. 15.
71. Boyle, *Things above Reason, Works,* vol. 4, pp. 424–445. Note that the text is mispaginated, skipping from p. 424 to p. 445. I have checked Birch's edition against the original, and Birch's version is complete.
72. Boyle, *Reflections upon a Theological Distinction, Works,* vol. 5, p. 547.
73. Boyle, *Excellency of Theology, Works,* vol. 4, p. 26.
74. Boyle, *Advices, Works,* vol. 4, p. 452.

At this point, I want to stress that Boyle's writings on the limits of human understanding would be of philosophical interest even if those writings were confined to the context of theology. In the seventeenth century, the fields of theology and philosophy (or natural philosophy) were not sharply delineated, and a strong argument could be made that lacking any evidence to the contrary, Boyle's views on the limits of human understanding in the context of theology should be construed as representative of his views on the limits of human understanding in general. Boyle, however, eliminated the need for such an argument; although it is clear from the preceding chapters that he organized and published his views on the limits of human understanding because of particular controversies that occurred in the context of theology, he explicitly designated his *Discourse of Things above Reason* as a *philosophical* and not a *theological* work,[75] and in it he interwove examples of things transcending reason in natural philosophy with the examples chosen from theology.

One reason Boyle might have had for designating *Things above Reason* as a philosophical work is that by doing so, he could state his opinions on the theological controversies (concerning things above reason in general and concerning predestination in particular) that had preceded *Things above Reason,* but without explicitly entering into those theological controversies himself. Another and more compelling reason for designating *Things above Reason* as a philosophical work, however, is that Boyle did indeed believe that human understanding is limited where the secrets of nature are concerned, just as it is limited where the mysteries of Christianity are concerned. There is abundant evidence of this not only in Boyle's writings concerning things above reason, but also in those of his works more directly concerned with questions of natural philosophy. In Part II, I discuss Boyle's views on things above reason in the context of natural philosophy, and the ways in which those views influenced his conception of the proper task of the natural philosopher.

75. Boyle, *Things above Reason, Works,* vol. 4, pp. 413, 418. However, the work was classified by his publishers and by compilers of his works as theological in nature.

PART II

The Context of Natural Philosophy

5

Philosophies of Nature and their Theological Implications

In Part II, I examine the relationship between Boyle's views on the limits of human reason and his conception of the task of the natural philosopher. The fact that I am now switching from an explicit discussion of theology in Boyle's thought to a discussion of his natural philosophy does not by any means involve leaving theological considerations behind; indeed, it is the central thesis of this book that Boyle's theological beliefs significantly influenced his view of the proper task of the natural philosopher. In many passages of various so-called "scientific" works, Boyle made his theological concerns quite clear, and when he did so I will be quick to note them. At other times, however, he seemed to think that he could isolate his concerns as a natural philosopher from his concerns as a lay theologian. This is evident in the fact that on numerous occasions he identified himself as speaking not as a Christian virtuoso, but as a natural philosopher who was excluding theological considerations from a given discussion. In *A Disquisition about Final Causes,* for example, he ended one discussion by stating that he had been speaking "meerly upon physical grounds," and then introduced his next point by announcing that "if the revelations contained in the holy scriptures" might be admitted he could offer other arguments concerning God's ends in nature.[1]

If Boyle's theological views and his conception of the proper task of the natural philosopher were inseparable, as I claim, why did he sometimes present himself as thinking that they could be separated? It is possible, of course, that he believed that he *could* separate the two, although in retrospect it is all too obvious that he failed to do so more often than not. A more likely explanation, however, is that he was acutely aware that his theological views and his corpuscularian matter theory constituted a seamless worldview, and that he was *also* acutely aware that the alternative theories of matter espoused by some of his

1. Boyle, *Final Causes, Works,* vol. 5, p. 411; see also *Notion of Nature, Works,* vol. 5, p. 189, and *Things above Reason, Works,* vol. 4, p. 423.

contemporaries were influenced by their *own* theological concerns. In short, the audience Boyle was writing for consisted of diverse groups of contemporaries, and the members of each group had their own theological presuppositions inextricably intertwined with their own preferred matter theory. In addition to the "libertines" and "atheists" to whom some of his works were addressed (and presumably he included in this category the followers of Hobbes and Spinoza), his readers included Aristotelians, the Chemical philosophers, and the Cambridge Platonists.

Clearly it was appropriate for Boyle to exclude theological considerations from passages aimed at "libertines" and "atheists," for obviously such readers (if indeed, there *were* any such readers) were not likely to be persuaded by arguments based on theological considerations.[2] That Boyle was aware of this is illustrated by the fact that in *The Excellency of Theology,* for example, he noted that because he was *not* arguing against atheists in that work, he felt free to include in it arguments drawn from theology.[3] What is less clear (and what I want to stress) is that it was also appropriate for Boyle to be very cautious in associating the corpuscularian philosophy with theological considerations in works targeted at Aristotelians, the Chemical philosophers, and the Cambridge Platonists. Why? Because the proponents of each of these alternative philosophies of nature were themselves Christians – Christians who had their own theological motives for accepting the various natural philosophies that they did, and Boyle was reluctant to trample on his contemporaries' religious sensibilities. Sometimes, when theology and natural philosophy were treated in very general terms (as they were in *Excellency of Theology,* for example), Boyle felt free to speak as a Christian to his Christian contemporaries. At other times, when he was championing the corpuscular hypothesis specifically in opposition to the matter theories held by his fellow Christians, he saw fit to (at least attempt to) divorce theology from natural philosophy in order to keep the area of dispute as narrow as possible. For example, in *Origin of Forms and Qualities,* he explicitly noted that many of the arguments of the Schoolmen were

> either confessedly, or at least really built upon some theological *tenets* of theirs, which being opposed by the divines of other churches, and not left unquestioned by some acute ones of their own, would not be proper to be solemnly taken notice of by me,

---

2. On the perceived problem of atheism, see the sources cited in chapter 3, note 5.
3. Boyle, *Excellency of Theology, Works,* vol. 4, p. 4.

whose business in this tract is to discourse of natural things as a naturalist, without invading the province of divines.[4]

In the chapters that follow, I examine in detail the relationship between Boyle's theological views and his own advocacy of the corpuscularian philosophy. In the remainder of this chapter, I examine the philosophies of nature of the Aristotelians, the Chemical philosophers, and the Cambridge Platonists, emphasizing in each case examples of some of the theological motivations of the proponents of each of these alternative philosophies of nature. Theological motivations were an essential and inseparable part of the matter theories of the seventeenth century, and recognizing this fact makes it easier to understand why Boyle treated certain topics as gingerly as he did.

Such an overview of these philosophies will necessarily be superficial. Each of these systems was far too complex to be adequately summarized in a few brief paragraphs, and the problem is compounded by the fact that there were significant variations of belief and emphasis among the many proponents of each philosophy. It should be remembered that my purpose here is to offer examples of the interrelatedness between theories of matter and theological concerns and not an exhaustive description of those relationships.

The Aristotelians

First there was the hylomorphism of the "vulgarly received notion of nature" of the Schoolmen, the philosophers working within the Aristotelian tradition. The Aristotelianism of the Renaissance was an extraordinarily diverse tradition; here I discuss it in broad outline only, with an emphasis on those aspects that were of particular interest to Boyle.[5]

---

4. Boyle, *Origin of Forms and Qualities*, vol. 3, p. 7. For Boyle's caution in dealing with fellow Christians, see Edward B. Davis, " 'Parcere nominibus': Boyle, Hooke and the Rhetorical Interpretation of Descartes," in *Robert Boyle Reconsidered*, edited by Michael Hunter (Cambridge: Cambridge University Press, 1994), pp. 157–175.

5. For diversity in the Aristotelian tradition, see Charles B. Schmitt, *Aristotle and the Renaissance* (Cambridge: Harvard University Press, 1983). In my discussion of Scholasticism, I have relied heavily on a number of sources, including Étienne Gilson, *History of Christian Philosophy in the Middle Ages* (New York: Random House, 1955); M.A. Stewart, Introduction to *Selected Philosophical Papers of Robert Boyle* (Manchester University Press,

The Schoolmen believed all bodies to be composed of a mixture of the four elements (earth, water, air, and fire), conditioned by the four qualities (hot, cold, wet, and dry) – a mixture that did not exist per se, but that was *informed*. Forms made each body the kind of object it is; without form, matter was believed to be a pure conceptualization without ontological reality. What actually exists, according to the Schoolmen, is individual objects, composites of matter and form.

Form, then, was what accounted for the properties of an object, and the Schoolmen attributed more than one form to natural objects. The *substantial* form accounted for the essence, or nature, of the object; it accounted for the characteristic behavior of that *kind* of substance (human beings reason, horses neigh, fire heats). Accidental forms accounted for properties that might vary among individuals of the same essential nature (some horses are black, some are white).

Substantial forms were invoked to explain the natural motions of objects, which were hence endowed with their own source or cause of activity, and this activity was conceived as each body's attempt to fulfill its own nature, an activity that could easily be conceived as the having of intentions and the exercising of volition. When generalized, this conception of purposeful activity resulted in the reification of nature as a whole (as "Nature," or, as Boyle put it, as an "almost divine thing")[6] invoked to explain natural phenomena in general terms. "Nature," it was claimed, was an entity that, for example, "abhors" a vacuum and always acts to prevent one.

The Schoolmen had their own theological reasons for insisting on the truth of Aristotelianism as it had developed in the context of Christianity. In the particular context of Roman Catholic theology, for example, the distinction between essence and accident had provided an explanation of the miracle of Transubstantiation, specifically that at the Consecration, God miraculously changes the essence of the bread and wine

---

1979; paperback edition, Indianapolis, IN: Hackett Publishing Co., 1991); A. George Molland, "Aristotelian Science," in *Companion to the History of Modern Science*, edited by R.C. Olby, G.N. Cantor, J.R.R. Christie, and M.J.S. Hodge (London and New York: Routledge, 1990), pp. 557–567; Daniel Garber, *Descartes' Metaphysical Physics* (Chicago: University of Chicago Press, 1992), pp. 94–103; Margaret J. Osler, *Divine Will and the Mechanical Philosophy*, pp. 171–179; and Charles H. Lohr, "Metaphysics," in *The Cambridge History of Renaissance Philosophy*, edited by Charles B. Schmitt, Quentin Skinner, Eckhard Kessler, and Jill Kraye (Cambridge: Cambridge University Press, 1988), pp. 537–638.

6. Boyle, *Vulgarly Received Notion of Nature, Works*, vol. 5, p. 161.

literally into the body and blood of Christ, leaving the accidents of taste and appearance unchanged.[7]

The Schoolmen also invoked the conception of a substantial form to explain (in a variety of ways) the relationship between the human body and the human soul. In doing so, they were acutely aware of the implications of any given interpretation of this relationship for the immortality of each "person." I use the word "person" advisedly, for the Christian doctrine of immortality is that of the resurrection of the body, a doctrine that historically has left the Christian theologian between a rock and a hard place over the relationship between body and soul. Too sharp a soul-body dualism might allow for the immortality of the soul, but might also leave the body behind to decay, a result that was theologically unacceptable, for if the soul should exist independently of the body, what assurance would there be that such a soul would be an individual retaining any knowledge of the sensible world? On the other hand, if soul and body were *not* distinct entities, the possibility that the soul perishes with the body reared its ugly head.

The story of the sheer variety of explanations of the soul-body relationship forwarded from the time of the early Church fathers on throughout the Middle Ages and Renaissance (and beyond) is a fascinating one, but not one that can be retold here.[8] Instead, I will present just one such explanation, one that had become fairly widely accepted by about 1330 – that of Thomas Aquinas.[9] By introducing the notion of *esse* as the act of the substantial form, Aquinas was able to preserve both the intimate union of soul and body *and* the separate subsistence of the human soul between the death of the body and its resurrection. After death, according to Aquinas, the soul remains a substance composed of its essence and its act of being, yet with the resurrection of the body the intimate union is restored.

Clearly, anyone who attacked the Scholastic account of substantial

---

7. For Descartes' difficulties with the Scholastics over the doctrine of Transubstantiation, see Richard A. Watson, "Transubstantiation among the Cartesians," in *Problems of Cartesianism,* edited by Thomas M. Lennon, John M. Nicholas, and John W. Davis (Kingston and Montreal: McGill-Queen's University Press, 1982), pp. 127–148. For Galileo's, see Pietro Redondi, *Galileo Heretic (Galileo Eretico)* (Princeton, NJ: Princeton University Press, 1987).

8. See Caroline Walker Bynum, *The Resurrection of the Body in Western Christianity, 200–1336* (New York: Columbia University Press, 1994).

9. Gilson, *History of Christian Philosophy in the Middle Ages,* pp. 376–377, 416–420; Bynum, *The Resurrection of the Body,* esp. pp. 256–277.

forms was also tampering with a centuries-long struggle to account for personal immortality in terms of hylomorphism (that is, of matter and form); to eliminate substantial forms was, from the point of view of the Schoolmen, to eliminate personal immortality as well. Boyle was aware of the theological implications the Schoolmen would read into any denial that the human soul is the substantial form of the human body. In *Origin of Forms and Qualities,* for example, a work devoted to the elimination of substantial forms from natural philosophy, he noted that the arguments of many modern Schoolmen were based on "some theological tenets of theirs," and so he explicitly exempted the human soul from his arguments against substantial forms.[10] He also exempted the human soul from his critique of substantial forms in *Notion of Nature.*[11] Boyle himself was able to avoid the pitfalls involved in these various conceptions of the human soul by appealing to the limits of reason; he included the nature of the union of the human soul to its body in his category of inexplicable things.[12]

The Cambridge Platonists

The second philosophy of nature to be considered – that propounded by the Cambridge Platonists – had much in common with the Scholastic philosophy in that it postulated an intermediary between God and natural phenomena, although the entity was a "hylarchic principle" or "plastic nature" rather than the substantial forms of the Schoolmen. Henry More's initial acceptance of and ultimate rejection of Cartesian mechanism has received a great deal of scholarly attention; clearly, in ultimately rejecting mechanism More had come to believe that if the two principles of matter and motion alone account for all natural

10. Boyle, *Origin of Forms and Qualities,* vol. 3, pp. 8, 12; quotation from p. 8.
11. Boyle, *Origin of Forms and Qualities, Works,* vol. 3, p. 40; *Notion of Nature, Works,* vol. 5, p. 166.
12. Boyle, *Things above Reason, Works,* vol. 4, pp. 413, 423; *Advices, Works,* vol. 4, p. 454; *Reason and Religion, Works,* vol. 4, p. 170. I discuss Boyle's treatment of substantial forms in more detail in chapter 7. Boyle's contemporary, Henry More, did struggle with the relationship of soul to body, and as a result was charged with being a materialist; see John Henry, "A Cambridge Platonist's Materialism: Henry More and the Concept of Soul," *Journal of the Warburg and Courtauld Institutes* 49 (1986), pp. 172–195.

phenomena, then there is no room for God's activity in the created world.[13]

Ralph Cudworth had similar concerns, and argued that a "plastick nature" exists and is the intermediary between God and the created world.[14] If there were no such intermediary, he thought, then either everything that happens is the result of chance or is the result of God's direct intervention in the created world. The former he rejected on the grounds that it is consistent with an irrational universe and atheism. He also rejected the idea that natural phenomena are the result of God's direct intervention in the created world. If God does everything directly, Cudworth explained, this would make him "immediately do all the Meanest and Triflingest things himself Drudgingly, without making use of any Inferior and Subordinate Instruments,"[15] and this would not be consistent with God's wisdom. Therefore there must be some "*Agent* and *Executioner*" that implements God's will in the world, and this agent is the "plastick nature," or "Substance Incorporeal."[16]

In addition to its role in confuting atheism, Cudworth's "plastick nature" also provided the solution to the problem of evil. Because it was incorporeal, it allowed for spiritual activity in the world, but because it

13. See, for example, Robert A. Greene, "Henry More and Robert Boyle on the Spirit of Nature," *Journal of the History of Ideas* 22 (1962), pp. 451–474; Alan Gabbey, "Henry More and the Limits of Mechanism," in *Henry More (1614–87): Tercentenary Studies,* edited by Sarah Hutton (Dordrecht: Kluwer Academic Publishers, 1990), pp. 19–35; idem, "Philosophia Cartesiana Triumphata: Henry More (1646–1671)," in *Problems of Cartesianism,* pp. 171–250; John Henry, "Henry More versus Robert Boyle: The Spirit of Nature and the Nature of Providence," in *Henry More,* pp. 55–76.

14. Cudworth's dualism involved a distinction between activity and passivity rather than mind and body; see John Passmore, *Ralph Cudworth: An Interpretation* (1951; reprint Bristol: Thoemmes, 1990), pp. 19–28.

15. Ralph Cudworth, *The True Intellectual System of the Universe: The First Part; Wherein, All the Reason and Philosophy of Atheism is Confuted; and its Impossibility Demonstrated* (Faksimile-Neudruck der Ausgabe von London 1678, Stuttgart-Badd Cannstatt: Friedrich Frommann Verlag [Günther Holzboog], 1964), p. 149.

16. Cudworth, *True Intellectual System,* pp. 147, 148. Cudworth rejected the possibility that "inferior spirits" (that is, "Dæmons or Angels") implement God's will in the creation, as well as the possibility that God's will should be understood as a concept of "*Verbal Law,*" the latter on the grounds that matter, being inanimate, cannot understand commands, and that such a conception would reduce God to an "*Idle Spectator*" (ibid., p. 148).

was unconscious and so deeply immersed in matter, it was responsible for the *"Errors* and *Bungles"* that are obvious in the natural world. According to Cudworth, if God did everything directly, his omnipotence would guarantee perfection in the created world.[17] The "plastick nature," on the other hand, lacking consciousness, "doth not at all Comprehend nor understand the Reason of what it self doth."[18] It implements God's will, but does so imperfectly, and remains subservient to "a Higher Providence ... which presiding over it, doth often supply the *Defects* of it, and sometimes Overrule it."[19]

I discuss Boyle's response to the "plastic nature" (or, in Henry More's terms, the "hylarchic principle"), as well as his response to the substantial forms of the Schoolmen in chapter 7. Here I simply note that Boyle had no difficulty with the idea that God can conserve and direct the created world from afar, even though he (Boyle) did not understand just how this could be. "The things, that are done in the corporeal world, are really done by the parts of the universal matter, acting and suffering according to the laws of motion established by the Author of nature," he claimed, going on to argue that although

> it is true, that it is not easy to conceive, how one agent should, by so simple an instrument as local motion, be able to direct a multitude of agents, as numerous as the bodies that make up a world, to act as regularly, as if each of them acted upon its own particular design, and yet all of them conspired to obey the laws of nature. But if we consider, that it is to God, that is an omniscient and almighty Agent, that this great work is ascribed, we shall not think it incredible.[20]

Both Boyle and Cudworth believed that God occasionally intervenes directly in nature. The difference between them was that Cudworth

17.   Cudworth, *True Intellectual System,* p. 150.
18.   Cudworth, *True Intellectual System,* p. 155.
19.   Cudworth, *True Intellectual System,* p. 150. See also the excellent discussion by Gerald R. Cragg, editor, *The Cambridge Platonists* (New York: Oxford University Press, 1968), pp. 234–235. An explication of Boyle's own solution to the problem of evil lies outside the scope of this discussion, other than to note that he appealed to reason's limits. There is evidence, however, that when Boyle made his comments about evil in *Final Causes* (*Works,* vol. 5, pp. 422–423) Cudworth was one of the people he had in mind; see Davis, "'Parcere nominibus'," p. 163, but note also Davis's *caveat,* p. 173 n. 34.
20.   Boyle, *Final Causes, Works,* vol. 5, pp. 409, 414. This passage is equally applicable to the Scholastic conception of substantial forms, or of "Nature" as an entity intermediate between God and natural phenomena.

believed that God ordinarily implements his will by means of an intermediary, whereas Boyle simply did not know how God directs nature on a
day-to-day basis. In Boyle's view, the claim of the Cambridge Platonists
to have understood such matters reflected their failure to understand
reason's limits, and their unwillingness to countenance the possibility
(in Boyle's view, the likelihood) that God in some mysterious way is
immediately responsible for the daily course of nature reflected their
failure to appreciate fully God's power.

The "Chymical" Tradition

Boyle was portrayed in the not-so-distant past as the natural philosopher
who definitively distinguished modern chemistry from its alchemical
origins.[21] More recently, however, it has become clear not only that
Boyle failed to make this distinction, but also that he himself was deeply
involved in such archtypal alchemical pursuits as the transmutation of
metals and the search for the Philosopher's Stone. Unfortunately, our
present understanding of Boyle-as-alchemist is fragmentary. This situation will be rectified by Lawrence M. Principe, who, in his much-needed
book-length exposition *The Aspiring Adept: Robert Boyle and his Alchemical Quest*, traces Boyle's involvement with *chrysopoeia* (gold-
making) over the course of his lifetime and provides a comprehensive
treatment of Boyle's relationship to the "chymical" tradition.[22]

   In this section, I adopt the term "chymical" as an all-encompassing
term to refer to the totality of chemistry and alchemy in Boyle's time. In
doing so, I follow the example set by Principe, who argues clearly and
convincingly that Boyle and most of his contemporaries used the terms
"chemistry" and "alchemy" interchangeably in referring to a domain
that was then not terminologically differentiated as it is today.[23] When I

21. Marie Boas [Hall], *Robert Boyle and Seventeenth-Century Chemistry*
    (Cambridge: Cambridge University Press, 1958).
22. I am grateful to Principe for having allowed me to read drafts of chapters
    of this important work as they have been completed; my discussion of the
    chymical tradition in this section owes much to his exposition.
23. Lawrence M. Principe, *Aspiring Adept: Robert Boyle and his Alchemical
    Quest*, Introduction (Princeton: Princeton University Press, forthcoming).
    William H. Newman makes the same point in *Gehennical Fire: The Lives
    of George Starkey, an American Alchemist in the Scientific Revolution*
    (Cambridge: Harvard University Press, 1994), pp. xi–xii. Hereafter, I elimi-
    nate quotation marks in references to chymistry.

use the term "alchemy," I mean it to reflect the distinction we in the twentieth century make between chemistry, accepted as a modern science, and alchemy, understood as a miscellany of archaic pursuits concerned with such activities as attempts to transmute base metals into gold and silver.

There was a tremendous flourishing of Paracelsian and other chymical works during the 1640s and 1650s in England, a flourishing that has been linked to the reform programs of the Puritans.[24] This burst of activity reflected the widespread interest in the so-called "occult sciences" during the sixteenth and seventeenth centuries, a subject that has received much scholarly attention in recent years. Although in many of the earlier of these studies, emphasis was placed on the Hermetic tradition, scholars have now begun to emphasize the extent to which Neoplatonic, Stoic, and Gnostic metaphysics and epistemology provided the conceptual frameworks to accommodate the natural magic of the Renaissance as expressed in astrology, alchemy, numerology, and related arts.[25] In this section, I limit myself to a discussion of the chymists'

24. Charles Webster, *The Great Instauration: Science, Medicine and Reform, 1626–1660* (London: Duckworth, 1975), pp. 273–282; P. M. Rattansi, "Paracelsus and the Puritan Revolution," *Ambix* 11 (1963), pp. 24–32.
25. An early, and seminal, work on the Hermetic tradition was Frances A. Yates, *Giordano Bruno and the Hermetic Tradition* (Chicago: University of Chicago Press, 1964). More recent and sophisticated introductions to the relationship of magic to the development of science include Brian P. Copenhaver, "Astrology and Magic," in *The Cambridge History of Renaissance Philosophy*, Charles B. Schmitt, general editor (Cambridge: Cambridge University Press, 1988), pp. 264–300, and John Henry, "Magic and Science in the Sixteenth and Seventeenth Centuries," in *Companion to the History of Modern Science*, pp. 583–596. Bert Hansen, "Science and Magic," in *Science in the Middle Ages*, edited by David C. Lindberg (Chicago: University of Chicago Press, 1978), pp. 483–506, provides a valuable overview of the conceptual framework of natural magic. A brief and accessible introduction to the history of alchemy through the sixteenth century can be found in Robert P. Multhauf, "The Science of Matter," in *Science in the Middle Ages*, pp. 369–390. Scholars who have emphasized the Neoplatonic, Stoic, and Christian Gnostic foundations of Renaissance magic include Charles B. Schmitt, "Reappraisals in Renaissance Science," in Charles B. Schmitt, *Studies in Renaissance Philosophy and Science* (London: Variorum Reprints, 1981), pp. 200–214, and Brian Copenhaver, "Hermes Trismegistus, Proclus, and a Philosophy of Magic," in *Hermeticism and the Renaissance: Intellectual History and the Occult in Early Modern Europe*, edited by Ingrid Merkel and Allen G. Debus (Washington, DC: The Folger Shakespeare Library, 1988), pp. 79–110.

matter theory (and some of its accompanying theological implications), for the chymists were the proponents of the third viable alternative to the corpuscular hypothesis and the subject of Boyle's *Sceptical Chymist* (1661).

Before going further, I should note that to speak in terms of a single "matter theory" associated with the chymical tradition is an error because there were a number of such theories in circulation during the sixteenth and seventeenth centuries; there was, as one scholar has put it, "no single body of knowledge or doctrine that might be called chemistry."[26] Here, however, it will be sufficient to discuss the two most basic theories – the *tria prima* most often connected with the Paracelsians, and the more traditional dyad of Mercury and Sulphur adhered to by many non-Paracelsian chymists – because an accurate understanding of Boyle's views on alchemy depends on the distinction between these two theories.

Pre-Paracelsian matter theory was quite limited in its application, and was based on the medieval Islamic Mercury-Sulphur theory, which was itself based on Aristotle's wet and dry exhalations of the earth that combined underground to produce stones, minerals, and metals. These exhalations were believed to condense into two intermediate substances (or principles), Mercury and Sulphur, which were believed to bear some analogy to common mercury and sulphur. Mercury and Sulphur, in turn, combined in varying degrees and in different conditions of purity to produce the metals; the belief that the inferior metals were some immature or unripe form of the noble metals was consistent, of course, with the belief that base metals could be transmuted into silver or gold. There is considerable confusion (often deliberate) in the traditional chymical texts as to the exact nature of Mercury and Sulphur, but when the terms were used of materials above the surface of the earth, they were often used not to denote particular bodies but rather as analogical terms expressing relationships such as active-passive, masculine-feminine, or coagulant-coagulated.[27]

The Paracelsian matter theory of the *tria prima* (with Salt being added as the third principle), although bearing certain resemblances to the traditional dyad, differed from it in very significant ways. The most notable of these is that whereas the dyad of traditional chymistry had been invoked to explain only the formation of metals and some minerals, the Paracelsians conceived the *tria prima* as the source of *all* natural

26. Betty Jo Teeter Dobbs, *The Foundations of Newton's Alchemy, or "The Hunting of the Greene Lyon"* (Cambridge: Cambridge University Press, 1975), p. 43.
27. I owe this explanation to Principe, *Aspiring Adept*, chapter 2.

The Context of Natural Philosophy

phenomena, both terrestrial and celestial.[28] The Paracelsians, followers of the Swiss-German physician and alchemist Paracelsus (1493–1541), influenced by the spiritual unrest that followed the Reformation, envisioned a new philosophy based on scripture and creation rather than on Greek (or any other) philosophy, and their matter theory was but one aspect of a thoroughly spiritualized worldview. For example, they interpreted the Mosaic account of the creation in alchemical terms; God, himself the greatest of all alchemists, had created all things from prime matter in what was essentially an alchemical process.[29] In their view, the world was filled with occult correspondences and spiritual forces; further, they believed that all knowledge was divinely inspired, sent directly from God to a select few.

The obscurity of Paracelsus's works and the variety of interpretations of those works make any simple characterization of the Paracelsian matter theory problematic. The basic thesis, however, was that God had created prime matter from which the three principles (the *tria prima,* Mercury, Sulphur, and Salt) emerged, and the three principles were the source of the Aristotelian four elements (earth, water, air, and fire). A variation of this was the theory of Jan Baptista van Helmont (1577–1644), who held that water was the prime substratum from which, by means of processes of fermentation and by being informed by various seminal causes (latent organizing principles capable of inducing form into matter), all things had been created and to which all things would return. Many of Boyle's early experiments show clearly the influence of Helmontian ideas; most likely, his partial acceptance of those ideas in the 1650s reflects Helmont's having presented his theory as being consistent with the account of creation given in Genesis.[30]

28. This important last point is emphasized in Principe, *Aspiring Adept,* chapter 2.

29. Betty Jo Teeter Dobbs, *The Foundations of Newton's Alchemy,* pp. 40–43; 80–81; idem, *The Janus Faces of Genius: The Role of Alchemy in Newton's Thought* (Cambridge: Cambridge University Press, 1991), pp. 53–88; Allen G. Debus, "Fire Analysis and the Elements in the Sixteenth and the Seventeenth Centuries," *Annals of Science* 23 (1967), p. 128, reprinted (with same pagination) as Essay VII in Debus, *Chemistry, Alchemy and the New Philosophy 1550–1700: Studies in the History and Science of Medicine* (London: Variorum Reprints, 1987).

30. See Boyle's comments in *Sceptical Chymist, Works,* vol. 1, pp. 494–498. See also Antonio Clericuzio, "From van Helmont to Boyle: A Study of the Transmission of Helmontian Chemical and Medical Theories in Seventeenth-Century England," *British Journal for the History of Science* 26 (1993), pp. 303–334. For Helmont's account of creation, see Dobbs,

The relationship of Boyle's corpuscularianism to the chymical tradition is complex, and only recently have scholars begun to interpret that relationship in a way that makes sense of Boyle's own endeavors in what today is considered to be alchemy. His *Sceptical Chymist* has long been deemed to have constituted a historical break between what most historians have considered to be the occult pseudoscience of alchemy and modern chemistry.[31] It has now become clear, however, that his attack on the "vulgar spagyrists" and "generality of alchemists" in that work targets particular groups of chymists and is not at all a blanket condemnation of alchemistry.[32] Instead, Boyle's targets were those "lower order" chymists, some of whom were mere technicians whose practice was virtually devoid of any theory (for example, distillers, apothecaries), and others of whom were "cheats," who had given chymistry a bad name. Yet a third group of "lower order" chymists were the textbook writers, especially those within the Paracelsian tradition, who had extended the dyad of Mercury and Sulphur (responsible for the formation and constitution of metals and minerals) to the triad of Mercury, Sulphur, and Salt (as explaining *all* natural phenomena); these chymists had erred by basing too general an account on too few (and inconclusive) experiments. Nowhere in the *Sceptical Chymist* is there an attack on genuine *adepti*, the "higher order" chymists whose efforts at transmutation constitute most of what we today term alchemy. When the *Sceptical Chymist* is understood in this way, sense is made of Boyle's own preoccupation with gold-making, a preoccupation that seems to have become more intense over the course of his lifetime.[33]

Further, our understanding of the relationship of Boyle's corpuscularianism to chymical matter theory is facilitated by realizing that cor-

---

The Janus Faces of Genius, p. 49; Eugene M. Klaaren, *Religious Origins of Modern Science: Belief in Creation in Seventeenth-Century Thought* (Grand Rapids, MI: William B. Eerdmans, 1977), pp. 53–83. For Boyle's distancing himself from Helmontian views, see Michael Hunter and Edward B. Davis, "The Making of Robert Boyle's *Free Enquiry into the Vulgarly Receiv'd Notion of Nature* (1686)," *Early Science and Medicine* I (1996), pp. 204–271.

31. Marie Boas [Hall], *Robert Boyle and Seventeenth-Century Chemistry.*

32. Principe, *Aspiring Adept,* chapter 2. As Principe points out, the terms in quotations come from the two title pages of the first edition of the *Sceptical Chymists*; the page identifying "vulgar spagyrists" was the one reused for the second edition of 1680. See also Antonio Clericuzio, "Carneades and the Chemists," in *Robert Boyle Reconsidered,* pp. 79–90.

33. Principe, *Aspiring Adept,* chapter 2; see also idem, "Boyle's Alchemical Pursuits," in *Robert Boyle Reconsidered,* pp. 91–105.

puscularian matter theories were widely (although not universally) accepted by chymists in the seventeenth century and had respectable roots in medieval thought. The matter theory of the *Summa perfectionis* (written around 1310 and falsely attributed to the alchemist known in Latin as "Geber"), for example, rested on a distinction between the sizes of particles, the small (*pars subtilis*) and the large (*pars grossus*), a theory that accounted for transmutation in corpuscular terms and a theory that both influenced the formation of and was consistent with Boyle's own corpuscularianism.[34]

Of the three matter theories considered in this chapter – those of the Aristotelians, the Cambridge Platonists, and the chymists – it is clear that Boyle's own theory was most akin to the corpuscularian chymical one. Further, his corpuscularianism was consistent with the *tria prima* of the Paracelsians, for mercury, sulphur, and salt (considered as particles of matter rather than as the principles Mercury, and so forth) could be composed of corpuscles, although it should be noted that Boyle did not accept the Paracelsian extension of what had been originally a narrow explanation of the formation of metals and minerals into a grandiose account of the elementary constituents of all natural objects. He also rejected the vitalism of the Paracelsian theory on the grounds that it infringed on what he considered to be the sphere of God's direct activity in the world.[35] I discuss the relationship of Boyle's corpuscularianism to the *tria prima* in more detail in chapter 7.

Of the three matter theories discussed in this chapter, the more esoteric aspects of the chymical philosophy came the closest to bridging the gap between natural and supernatural phenomena. This caused Boyle considerable concern, and what we now know about his preoccupation with traditional alchemy raises a number of questions related to his theological concerns as well as questions related to the image of Boyle portrayed by scholars who have not taken this preoccupation into consideration in their studies of his natural philosophy. For example, although he has been depicted as an advocate of openness of communication among natural philosophers,[36] it is now clear not only that he

34. William R. Newman, "Boyle's Debt to Corpuscular Alchemy," in *Robert Boyle Reconsidered*, pp. 107–118.

35. Antonio Clericuzio, "A Redefinition of Boyle's Chemistry and Corpuscular Philosophy," *Annals of Science* 47 (1990), pp. 561–589; Barbara Beigun Kaplan, "*Divulging of Useful Truths in Physick: The Medical Agenda of Robert Boyle*" (Baltimore: The Johns Hopkins University Press, 1993), p. 60.

36. See, for example, M. E. Rowbottom, "The Earliest Published Writing of

devised elaborate techniques of concealment in order to protect his alchemical laboratory notes from prying eyes, but also that he utilized specifically alchemical techniques to disguise his attempts to make contact with *adepti* in some of his published works. One of the reasons for this secrecy was the belief of many *adepti* that alchemical knowledge was sacred.[37] Further, it was widely believed that spiritual entities might well be involved in the changes made in matter during such mysterious activities as the transmutation of base metals into gold.

I discuss Boyle's views on the possibility of a natural philosopher's being aided in his search for knowledge by spiritual beings in some detail in chapter 6; here it is enough to note that activities involving spirit-contact carry with them the possibility that the spirits with which one is involved might be the malevolent emissaries of the Devil rather than angels sent by God.[38] Boyle's views on the exact relationship of alchemy to theology are as yet not clear, nor is it yet clear to what extent his alchemical practice involved exceptions to the views he expressed when treating of nonalchemical topics. Principe's exposition in *Aspiring Adept* will provide the first comprehensive account of this intriguing topic.

In this chapter, I have provided a brief overview of the three main matter theories that provided the most viable alternatives to Boyle's corpuscular hypothesis. In doing so, I have emphasized some of the theological implications of each. This survey has necessarily been extremely selective, and the amount of material omitted is greater than the amount included. I hope, however, that I have included enough examples of the theological considerations of some of Boyle's contemporaries to make

Robert Boyle," *Annals of Science* 6 (1950), pp. 376–389. Most recently, Boyle's openness has been stressed by Kaplan, "*Divulging of Useful Truths in Physick*," esp. pp. 24–25. For a more accurate assessment of Boyle's views on openness in iatrochemical matters, see Michael Hunter, "The Reluctant Philanthropist: Robert Boyle and the 'Communication of Secrets and Receits in Physick'," in *Religio Medici*, edited by O. P. Grell and A. Cunningham (Aldershot: Scholar Press, 1996). For a more accurate view of Boyle's role in the controversy between Galenists and the iatrochemists than that provided by Kaplan, see idem, "Boyle versus the Galenists: A Suppressed Critique of Seventeenth-century Medical Practice and its Significance." *Medical History* 41 (1997).

37. Lawrence M. Principe, "Boyle's Alchemical Secrecy: Codes, Ciphers and Concealments," *Ambix* 39 (1992), pp. 63–74.

38. Michael Hunter, "Alchemy, Magic and Moralism in the Thought of Robert Boyle," *British Journal for the History of Science* 23 (1990), pp. 387–410.

my intended point – that in the seventeenth century, natural philosophy was intertwined inextricably with theology.

To attempt to deal with the natural philosophy of the period in isolation is at best misguided, and at worst, misleading. Only when we understand that Boyle's beliefs about the nature of the world and what can be known about it were influenced by his theological beliefs can we understand correctly the relationship between his ontology and his epistemology. Further, our understanding of his critiques of the matter theories of the Aristotelians, Cambridge Platonists, and Paracelsians is enhanced when we realize that both he and his readers were aware not only that philosophical points were being debated but that alternative interpretations of the Christian religion were being addressed, implicitly if not explicitly, as well.

It is to these topics I turn in the next three chapters. In chapter 6, I examine Boyle's evaluations of various sources of knowledge. In chapter 7, I explain the relevance of his *Discourse of Things above Reason* to his conception of the task of the natural philosopher. And in chapter 8, I show that his views on reason's limits were an integral part of his voluntaristic conception of the Judeo-Christian God.

6

Sources of Knowledge

I do not cover all aspects of Boyle's empirical epistemology in detail in this book because there have already been a number of excellent studies on the topic.[1] In this chapter, therefore, I limit myself to a discussion of the various ways in which Boyle believed he could attain knowledge of the created world. Specifically, I discuss Boyle's views on scriptural revelation, personal revelation, abstract reason (including innate ideas, which, having their origin in God, might be construed as a form of revelation), and sensory perception as possible sources of knowledge about the created world.

Scriptural Revelation

In *Defence of the Doctrine Touching the Spring and Weight of the Air ... Against the Objections of Franciscus Linus* (1662), Boyle stated that there are only two ways of obtaining knowledge of the created world:

1. The best work by far is Rose-Mary Sargent, *The Diffident Naturalist: Robert Boyle and the Experimental Philosophy* (Chicago: University of Chicago Press, 1995), although I should note that Sargent thinks it is misleading to characterize Boyle as an empiricist (pp. 2–14) and would therefore object to having her work characterized as a study of his empirical methodology. See also idem, "Learning from Experience: Boyle's Construction of an Experimental Philosophy," in *Robert Boyle Reconsidered*, edited by Michael Hunter (Cambridge: Cambridge University Press, 1994); Henry G. van Leeuwen, *The Problem of Certainty in English Thought, 1630–1690* (The Hague: Martinus Nijhoff, 1963), especially Richard H. Popkin's introduction to this work; Barbara Shapiro, *Probability and Certainty in Seventeenth-Century England* (Princeton: Princeton University Press, 1983); R.G. Frank, *Harvey and the Oxford Physiologists* (Berkeley and Los Angeles: University of California Press, 1980); Steven Shapin and Simon Schaffer, *Leviathan and the Air-Pump: Hobbes, Boyle, and the Experimental Life* (Princeton: Princeton University Press, 1985).

sense perception and revelation. Linus had attempted an explication of the "rota aristotelica" in which he appealed (among other things) to the motion of angels.[2] To this Boyle, replied that

> as for ... angels, having no immediate revelation, and a spirit and its actions not falling under sense, and not having any third way by which to be informed, I shall leave him [Linus] there to enjoy his fancies.[3]

It is not clear whether by "immediate revelation" about the actions of angels Boyle meant personal revelation or scriptural revelation. Because he viewed these as two distinct sources of knowledge, I examine them separately, beginning with scriptural revelation.

Boyle believed that scripture revealed knowledge of the created world only about matters that are not knowable by other means, such as the origin and ultimate destiny of the world. In *Usefulness of Experimental Natural Philosophy* (1663), he explained why it is, in his opinion, wise to rely on the account in Genesis, but unwise to rely on other passages of the Bible for knowledge of the natural world:

> For first [God] begins the book of scripture with the description of the book of nature, of which he not only gives us a general account, to inform us, that he made the world (since for that end, the very first verse in the bible might have sufficed) but he vouchsafes us by retail[ing] the narrative of each day's proceedings; and in the two first chapters of *Genesis*, is pleased to give nobler hints of natural philosophy than men are yet perhaps aware of. Though, that in

---

2. Born in England of Roman Catholic parents, Line was educated on the Continent and taught Hebrew and mathematics at Liège as a Jesuit priest. In the late 1650s, he was transferred to England where he worked until approximately 1665, primarily in London, to reconvert the English to Roman Catholicism, at which time he returned to Liège. While in England (where he was also known as Francis Hall and Thomas Hall; it is not certain exactly what his original name was), he became involved not only in the dispute with Boyle, but also in one with Isaac Newton. See Conor Reilly, S.J., *Francis Line S.J.: An Exiled English Scientist, 1595–1675*, Bibliotheca Instituti Historici S.I., vol. 29 (Rome: Institutum Historicum S.I., 1969). Boyle himself referred to the problem of the "rota Aristotelica" in *Things above Reason* (*Works*, vol. 4, p. 413). The question is why the circumference of an "inner" circle of a wheel traverses the same distance in the same time as the circumference of the wheel itself, which is a longer "line." The problem is found in *Mechanica* 24 855ᵃ28–856ᵃ38, a work falsely attributed to Aristotle. Boyle apparently did not know that Galileo had solved the problem. For a discussion of the problem, see Thomas Heath, *Mathematics in Aristotle* (Oxford: At the Clarendon Press, 1949), pp. 246–252.

3. Boyle, *Defence of the Doctrine, Works*, vol. 1, p. 183.

most other places of the scripture, where the works of nature are mentioned but incidently, or in order to other purposes, they are spoken of rather in a popular than accurate manner, I dare not peremptorily deny, being unwilling to interest the reputation of Holy Writ (designed to teach us rather divinity than philosophy) in the doubtful contentions of naturalists, about such matters, as may (though the history of the creation cannot) be known by the meer light of natural reason.[4]

Boyle repeated this point in *Excellency of Theology* (1674) when he noted that scripture was designed primarily "to teach us nobler and better truths, than those of philosophy," and stated that he did not agree with the

> opinion and practice [of those] that would deduce particular theorems of natural philosophy from this or that expression of a book, that seems rather designed to instruct us about spiritual than corporeal things.[5]

Although Boyle did not believe in relying on scripture for knowledge of particular theorems of natural philosophy that are concerned with the way the world presently is, in *Excellency* he did go on to discuss in some detail those things about the natural world about which he thought scripture *does* reveal truths that human beings cannot obtain in any other way, such as knowledge of the origin and age of the created world and the ultimate destiny of the created world and its contents. In fact, he claimed that where such truths are concerned, the limits of human reason are revealed by the errors of those who judge these matters by means of natural reason alone. Aristotle, for example, unenlightened by scriptural revelation, had reasoned that the world is eternal, and even the ancient philosophers who had taught that the world had a beginning had erred in thinking that matter is eternal. None of the ancient philosophers believed that any substance could be created out of nothing, yet, Boyle pointed out, scripture informs us that God indeed created the world from nothing; this is a truth that reason, unaided by revelation, could not reach. Since Aristotle's time, he noted, some philosophers have considered the world to be created but have been unable to determine how old the world is. Although some "ambitious people, such as those fabulous Chaldeans," have claimed that the world is some 40,000 or 50,000 years old, the truth (or so Boyle claimed) is that the world is only some 10,000 years old, and this truth is obtainable only by studying scripture.[6]

---

4. Boyle, *Usefulness of Experimental Natural Philosophy, Works,* vol. 2, p. 19.
5. Boyle, *Excellency of Theology, Works,* vol. 4, p. 11.
6. Boyle, *Excellency of Theology, Works,* vol. 4, pp. 10–11.

Further, Boyle argued that scripture is the only means we have of knowing those facts about creation that preceded the creation of the first man, and it is the only way of knowing the facts about the creation of Adam and Eve as well. Aware that some of his contemporaries reduced the first two chapters of Genesis to an allegory, Boyle himself would not reject the literal and historical sense of those chapters.[7] From scripture we know that God created Adam fully mature, not an infant (as some philosophers have fancied), and that although God created Adam out of terrestrial matter, he created Eve from Adam's rib. On the basis of this scriptural knowledge, we can reject the erroneous claims of those (such as the Epicureans) who attribute the origin of the human body to the casual concourse of atoms, as well as the claims of those (Boyle mentioned the Stoics and Hobbes) who argue that human beings sprang up "like mushrooms out of the ground."[8]

Knowledge of other things about the created world and its contents that, Boyle claimed, cannot be known other than by scriptural revelation (or that, at least, had never been ascertained by unassisted human reason) include the ultimate fate of the world itself, the resurrection of the human body, and the existence of angels. Although Boyle noted that Biblical revelation is not explicit about whether the world will one day be annihilated by fire or transfigured by fire, scripture does expressly reveal that the present frame of nature will be destroyed. Hence the opinion of those philosophers who hold that the duration of the world is indeterminable and has no certain limits can be rejected.[9] From scrip-

7. Boyle may have had Henry More's views on the allegorical interpretation of Genesis in mind; see Joseph Levine, "Latitudinarians, Neoplatonists, and the Ancient Wisdom," in *Philosophy, Science, and Religion in England 1640–1700*, edited by Richard Kroll, Richard Ashcraft, and Perez Zagorin (Cambridge: Cambridge University Press, 1992), pp. 85–108. For Edward Stillingfleet's objections to allegorical interpretations, see Sarah Hutton, "Edward Stillingfleet, Henry More, and the Decline of *Moses Atticus:* A Note on Seventeenth-century Anglican Apologetics," in the same volume, pp. 68–84. Somewhat later in the century, Thomas Burnet argued for an allegorical interpretation of Genesis in *Telluris Theoria Sacra* (1681 and 1689); for the ensuing controversy, see James E. Force, *William Whiston: Honest Newtonian* (Cambridge: Cambridge University Press, 1985), pp. 32–62.

8. Boyle, *Excellency of Theology, Works*, vol. 4, pp. 11–12; quote, p. 11. Boyle did not cite sources for his comments concerning the Epicureans, the Stoics, and Hobbes.

9. Boyle, *Excellency of Theology, Works*, vol. 4, p. 11. Boyle cited James, chapter 3, verse 6 ("And the tongue is a fire. The tongue is an unrighteous

ture, too, Boyle obtained knowledge of the resurrection of the body, as well as knowledge of the existence of angels.[10]

Personal Revelation

Scripture, for Boyle, was a valuable but limited source of knowledge of the created world; it was limited to the revelation of those truths about the world and its contents that could not be obtained by other means. Personal revelation, or knowledge obtained by spirit-contact, however, was another matter. This was a possibility Boyle took very seriously. It was also a possibility about which he had extremely ambiguous feelings because obtaining knowledge in this way involved the risk of contact with evil spirits.[11]

In *Usefulness of Experimental Natural Philosophy,* Boyle claimed that God might impart knowledge of nature to human beings by means of good angels, and in fact noted that there may be phenomena related to the natural world "which though we might understand well enough, if

---

world among our members, staining the whole body, setting on fire the cycle of nature, and set on fire by hell"), and 2 Peter, chapter 3, verse 7 ("But by the same word the heavens and earth that now exist have been stored up for fire, being kept until the day of judgment and destruction of ungodly men"), verse 10 ("But the day of the Lord will come like a thief, and then the heavens will pass away with a loud noise, and the elements will be dissolved with fire, and the earth and the works that are upon it will be burned up"), and verse 13 ("But according to his promise we wait for new heavens and a new earth in which righteousness dwells").

10. Boyle, *Excellency of Theology, Works,* vol. 4, pp. 9–11. Some Socinians were among those who believed that the resurrected body would be other than the earthly body; they argued that it would be a "glorified" or a "spiritual" body (see Joachim Stegmann, *Brevis Disquisitio: Or, a Brief Enquiry Touching a Better way Then is commonly made use of, to refute Papists, and reduce Protestants to certainty and Unity in Religion,* translated by John Biddle [London, 1653], pp. 29–32). In *Possibility of the Resurrection* (1675), Boyle argued that resurrecting the exact same physical body is not impossible for an omnipotent God. Boyle was not clear on whether natural reason can ascertain the immortality of the soul per se (compare *Christian Virtuoso, Works,* vol. 5, p. 517 with *Excellency of Theology, Works,* vol. 4, p. 14). Concerning angels, although their existence is revealed in scripture, Boyle was at times puzzled that Genesis contains no account of their creation; see *High Veneration, Works,* vol. 5, p. 148.

11. Michael Hunter, "Alchemy, Magic and Moralism in the Thought of Robert Boyle," *British Journal for the History of Science* 23 (1990), pp. 387–410.

God, or some more intelligent being than [we are], did make it his work to inform us of them, yet we should never of our selves find out those truths," and he cited scripture to confirm at least one such instance.[12] In the same work, however, he "dare[d] not affirm, with some of the Helmontians and Paracelsians, that God discloses to men the great mystery of chymistry by good angels, or by nocturnal visions."[13]

Nevertheless, Boyle did not go on in this work to recommend efforts to seek such supernatural information, exploring instead other ways in which he believed God had aided human beings in their search for natural knowledge. First, God may protect the attempts of naturalists to obtain knowledge by "protecting their attempts from those unlucky accidents, which often make ingenuous and industrious endeavours miscarry." Second, God might supernaturally foster friendships between one naturalist and another, thus facilitating the free exchange of information that one of the naturalists might have achieved as the result of many years' efforts and that might otherwise be kept secret. Third, the principal way in which God might increase man's knowledge of the natural world was by supernaturally directing the naturalist's attention to "those happy and pregnant hints" that, once considered, might lead to significant discoveries. Indeed, Boyle believed that "the two great inventions of the latter ages, gunpowder, and the loadstone's respect unto the poles," seemed to have been more the result of such (apparent) chance than "any extraordinary skill in philosophical principles."[14]

Spirit contact in seventeenth-century England was not the exclusive domain of "Helmontians and Paracelsians." Like his contemporary, Joseph Glanvill, Boyle recognized that empirical evidence of the existence of spirits would do much to counter the arguments of atheists. Boyle, of course, needed no empirical evidence of the spirit world because the existence of both good and bad angels is revealed in scripture.

12. Boyle, *Usefulness of Experimental Natural Philosophy, Works,* vol. 2, pp. 46, 61. Boyle cited Genesis 31, where God had taught Jacob how "to make lambs and kids come into the world speckled, and ring-streaked" (verses 8–11).

13. Boyle, *Usefulness of Experimental Natural Philosophy, Works,* vol. 2, p. 61. Boyle's early alchemical collaborator, George Starkey, had claimed in a letter to Boyle that alchemical knowledge had been imparted to him by an angel sent from God; see Lawrence M. Principe, "Boyle's Alchemical Pursuits," p. 100. On Starkey, see William R. Newman, *Gehennical Fire: The Lives of George Starkey, an American Alchemist in the Scientific Revolution* (Cambridge: Harvard University Press, 1994).

14. Boyle, *Usefulness of Experimental Natural Philosophy, Works,* vol. 2, p. 61.

But atheists reject scriptural revelation; hence Boyle wrote to Glanvill, encouraging him in his attempt to document accounts of the activities of spirits. If, out of the many "false or suspicious" accounts of spirits, even one verified account of "intelligent beings, that are not ordinarily visible," could be provided, then the atheists' rejections of such phenomena would be invalidated.[15]

Ultimately, however, despite Boyle's belief that knowledge of the natural world could be imparted by spirits, despite his belief that documented evidence of spiritual activity in the world would counter the arguments of atheists, and despite his own life-long pursuit of alchemical secrets, Boyle – at least at one point in his life – tentatively and reluctantly rejected attempts to contact the spirit world as illicit because doing so might bring one in contact with evil, as well as good, spirits. The investigation of spiritual substances other than human souls was, he thought, hazardous; the desire of "over-greedy" persons to investigate such spirits had caused them to put themselves "within the power of dæmons, [rather] than remain ignorant whether or not there are any such beings."[16] The problem, as Michael Hunter has put it, was that

> if you believed, as Boyle so strongly did, in an active supernatural realm, inhabited as much by the Devil and his attendant demons as by God and his angels, then this meant that there was a real possibility that you might come into contact with the demonic realm. . . . You would have to decide whether or not to pursue such contact on the grounds, not only of truth and falsehood, but also of what was licit and what was illicit.[17]

Clearly, Boyle was almost literally "of two minds" where the tantalizing possibility of achieving knowledge of nature's secrets via contact with spirits was concerned. Of more importance in this book than his mixed feelings about spirit-contact, however, is the fact that Boyle obviously thought that angels, both good and bad, have knowledge that is intermediate between the knowledge obtainable by human beings and God's complete and perfect knowledge.

15. Boyle to Glanvill, 10 February 1678, *Works*, vol. 6, pp. 57–59. For Glanvill's (and Henry More's) attempts to verify empirically the existence of spirits, see Jackson I. Cope, *Joseph Glanvill: Anglican Apologist* (St. Louis: Washington University Studies, 1956), pp. 87–103.

16. Boyle, *Excellency of Theology, Works*, vol. 4, p. 6.

17. Hunter, "Alchemy, Magic, and Moralism," p. 396. For a more recent study of Boyle's mixed feelings concerning spirit-contact, see Lawrence M. Principe, *The Aspiring Adept: Robert Boyle and his Alchemical Quest* (Princeton: Princeton University Press, forthcoming), chapter 6.

For to be above our reason is not an absolute thing, but a respective one, importing a relation to the measure of knowledge, that belongs to the human understanding, such as it is said to transcend: and therefore it may not be above reason, in reference to a more enlightened intellect, such as in probability may be found in rational beings of an higher order, such as are the angels, and, without peradventure, is to be found in God, whom when we conceive to be a Being infinitely perfect, we must ascribe to Him a perfect understanding and boundless knowledge.[18]

Abstract Reason and Innate Ideas

Given God's "perfect understanding and boundless knowledge," might he not have revealed some of that knowledge to human beings via the medium of innate ideas, thereby facilitating the acquisition of knowledge of the natural world by reason alone? It is impossible to determine Boyle's exact views on innate ideas from his scattered references to them. To my knowledge, Boyle denied the existence of innate ideas in only one passage, when he noted that the intellect of man is a bounded faculty, "naturally furnished with no greater a stock or share of knowledge, than it is able by its own endeavours to give itself, or acquire."[19] In other passages, he assumed their existence. Nevertheless, it is clear that he did not consider such ideas (if indeed there are any) to be a significant source of knowledge, whether that knowledge be of the created world or of theological truths.[20] Further, Boyle believed that even if human beings do have innate ideas, those ideas do not extend to all knowable objects but reach only as far as God intended they should. In *The Christian Virtuoso* (1690), for example, he considered the possibility that some individuals might think that his emphasis on experience degrades the human intellect. Boyle denied that he was degrading the faculty of reason, and insisted that reason is absolutely necessary to judge rightly of the information presented to it by experience, as well as to correct sensory perceptions that are misleading (such as the perception of a square tower that appears round from a distance, or of an oar partially

18. Boyle, *Reflections upon a Theological Distinction, Works,* vol. 5, p. 547.
19. Boyle, *Reflections upon a Theological Distinction, Works,* vol. 5, p. 545.
20. For controversies between rationalists and empiricists in the seventeenth century over the existence of and (if they exist) the role of innate ideas in acquiring knowledge, see Robert Merrihew Adams, "Where Do Our Ideas Come From? – Descartes vs. Locke," in *Innate Ideas,* edited by Stephen P. Stitch (Berkeley and Los Angeles: University of California Press, 1975).

immersed in water that appears crooked).[21] But "mere" or "abstracted"
reason (that is, reason divorced from experience) is not, in his opinion,
an adequate standard of truth. As Boyle put it, he was

> unapt to take, for the adequate standard of truth, a thing so imper-
> fectly informed, and narrowly limited, as his mere or abstracted
> reason, (as . . . one may call that, which is furnished only with its
> own, either congenite, or very easily and very early acquired notions
> and ideas, and with popular notices).[22]

Similarly, he observed that

> whether or not it be true . . . that the understanding is like blank
> paper; and that it receives no knowledge, but what has been con-
> veyed to it through the senses . . . it is plain, that the notions, which
> are either congenite with the understanding, or so easily and early
> acquired by it, that divers philosophers think them innate, are but
> very few in comparison of those, that are requisite to judge aright
> about . . . things, that occur either in natural philosophy, or the-
> ology.[23]

Further, in Boyle's view, those who rely on abstracted reason alone
(that is, without the assistance of experience and revelation) will form
very deficient, if not totally mistaken, ideas not only of God but also of
the world that God created:

> It must frequently happen, that the notions and opinions men take
> up of the works and mind of God, upon the mere suggestions of the
> abstracted reason, . . . must not only be almost always very defi-
> cient, but will be oftentimes very erroneous. Of which we see evi-
> dent proofs in many of the opinions of the old philosophers, who,
> though men of strong natural parts, were misled by what they
> mistook for reason, to maintain such things about the works and
> the author of nature, as we, who, by the favour of experience and
> revelation, stand[ing] in a much clearer light, know to be false, and
> often justly think utterly extravagant. . . . Abstracted reason, is but
> a narrow thing, and reaches but to a very small share of the multi-
> tude of things knowable, whether human or divine, that may be
> obtained by the help of further experience, and supernatural revela-
> tion. This reason, furnished with no other notices than it can supply
> itself with, is so narrow and deceitful a thing, that he, that seeks for
> knowledge only within himself, shall be sure to be quite ignorant of
> far the greatest part of things, and will scarce escape being mistaken
> about a good part of those he thinks he knows.[24]

---

21. Boyle, *Christian Virtuoso, Works,* vol. 5, p. 539.
22. Boyle, *Christian Virtuoso, Works,* vol. 5, p. 536.
23. Boyle, *Christian Virtuoso, Works,* vol. 5, p. 538.
24. Boyle, *Christian Virtuoso, Works,* vol. 5, p. 539.

Despite this obviously negative conception of the overall reliability of innate ideas, in those passages where he assumed their existence, Boyle also discussed their roles. First, they constitute knowledge that is recognized as being true as soon as it is proposed, such as that two contradictories cannot both be true.[25] Second, God has implanted "notions and principles in the mind of man fit to make him sensible, that he ought to adore God."[26] Even these two functions of innate ideas, however, are circumscribed by further considerations. If God has implanted such rules of reason as the principle of noncontradiction, God has also furnished man with a superior "innate light of the rational faculty" that is "more primary than the very rules of reasoning, since, by that light, we judge even" the limits of the principle of noncontradiction, from which some scriptural revelations may be exempt.[27] And even if God has indeed implanted in the human mind the notion of his existence and our duty to adore him, those philosophers who rely on innate ideas as a proof of God's existence (and Boyle specifically mentioned Descartes) detract from a superior argument concerning God's existence and man's duty – the contemplation of God's providence as exhibited in final causes.[28]

I reserve a discussion of Boyle's reasons for believing abstract reason to be such a limited and even misleading source of knowledge for chapter 8. For the moment, it is enough to note that pure reason, for Boyle, was definitely not a source of significant knowledge about the natural world.

Sensory Perception

The importance of sensory perception to an empirical scientific methodology is too well-known and too well-documented to require much comment. As Steven Shapin and Simon Schaffer have put it, "the experi-

---

25. Boyle, *Things above Reason, Works,* vol. 4, p. 414.
26. Boyle, *Christian Virtuoso, Works,* vol. 5, p. 520.
27. Boyle, *Advices, Works,* vol. 4, p. 460. I discussed this limitation in chapter 4.
28. Boyle, *Final Causes, Works,* vol. 5, p. 401. For Boyle's preference of the teleological argument for God's existence over the ontological argument, see Edward B. Davis, "'Parcere nominibus': Boyle, Hooke and the Rhetorical Interpretation of Descartes," in *Robert Boyle Reconsidered,* pp. 157–175, esp. p. 162. For an analysis of his argument, see Timothy Shanahan, "Teleological Reasoning in Boyle's *Disquisition about Final Causes,*" in the same volume, pp. 177–192.

mental philosophy, empiricist and inductivist, depended upon the generation of matters of fact that were objects of perceptual experience."[29] And as Boyle put it in *The Christian Virtuoso,*

> the philosophy, which is most in request among the modern virtuosi ... is built upon two foundations, reason and experience. But ... although the peripatetic (and some other) philosophies do also pretend to be grounded upon reason and experience; yet there is a great difference betwixt the use, that is made of these two principles, by the school-philosophers and by the virtuosi. For those, in the framing of their system, make but little use of experience. . . . But now, the virtuosi I speak of ... make a much greater and better use of experience in their philosophical researches. For they consult experience both frequently and heedfully; and ... they are careful to conform their opinions to it; or, if there be just cause, reform their opinions by it.[30]

I will not discuss further the primacy of sensory experience over reason in the thought of seventeenth-century empiricists, nor the significance of contrived experience (experiments) for the new natural philosophy. And I will not discuss the significance of the development and availability of the new instruments, such as the microscope, which enlarged the realm of the sensible for these empiricists – other than to note that although not all of Boyle's contemporaries agreed with him, he considered it likely that in time, corpuscles themselves would be perceivable by means of the microscope.[31] All of these topics have been covered so well elsewhere as to require no elaboration here.[32] Instead, I will discuss certain important points that Boyle made about objects which are not directly perceptible to the senses, whether those things be physical (corpuscles) or spiritual (spirits).[33]

29. Shapin and Schaffer, *Leviathan and the Air-Pump,* p. 36. See also R.G. Frank, *Harvey and the Oxford Physiologists.*
30. Boyle, *Christian Virtuoso, Works,* vol. 5, pp. 513–514.
31. Catherine Wilson, *The Invisible World: Early Modern Philosophy and the Invention of the Microscope* (Princeton: Princeton University Press, 1995), esp. pp. 51–60.
32. In addition to the works cited in note 1 of this chapter and Wilson, *Invisible World,* see Marie Boas [Hall], "The Establishment of the Mechanical Philosophy," *Osiris* 10 (1952), pp. 412–541; idem, "Boyle as Theoretical Scientist," *Isis* 41 (1950), pp. 261–268.
33. The inference from observable effects to the existence of insensible objects has been termed "transdiction"; see Maurice Mandlebaum, *Philosophy, Science and Sense Perception: Historical and Critical Studies* (Baltimore: The Johns Hopkins University Press, 1964). For a succinct explanation of transdiction in seventeenth-century natural philosophy, see Margaret J.

Boyle believed that a naturalist is justified in assuming that insensible material objects exist on the grounds of their observable effects. As he put it in *The Sceptical Chymist* (1661),

> we may well suspect, that there may be several sorts of bodies, which are not immediate objects of any one of our senses; since we see, that not only those little corpuscles, that issue out of the load-stone, and perform the wonders, for which it is justly admired; but the effluviums of amber, jet, and other electrical concretes, though by their effects upon the particular bodies disposed to receive their action, they seem to fall under the cognizance of our sight, yet do they not as electrical immediately affect any of our senses, as do the bodies, whether minute or greater, that we see, feel, taste, &c.[34]

Similarly, Boyle believed that human beings are justified in assuming God's existence on the grounds of observable phenomena:

> That the consideration of the vastness, beauty, and regular motions, of the heavenly bodies; the excellent structure of animals and plants; besides a multitude of other phænomena of nature, and the subserviency of most of these to man; may justly induce [man], as a rational creature, to conclude, that this vast, beautiful, orderly, and (in a word) many ways admirable system of things, that we call the world, was framed by an Author supremely powerful, wise, and good, can scarce be denied by an intelligent and unprejudiced con-

---

Osler, *Divine Will and the Mechanical Philosophy: Gassendi and Descartes on Contingency and Necessity in the Created World* (Cambridge: Cambridge University Press, 1994), p. 177. See also Peter Alexander, *Ideas, Qualities and Corpuscles: Locke and Boyle on the External World* (Cambridge: Cambridge University Press, 1985); F.J. O'Toole, "Qualities and Powers in the Corpuscular Philosophy of Robert Boyle," *Journal of the History of Philosophy* 12 (1974), pp. 295–315; and J.E. McGuire, "Atoms and the 'Analogy of Nature': Newton's Third Rule of Philosophizing," *Studies in History and Philosophy of Science* 1 (1970), pp. 3–58; McGuire uses the term "transduction" to refer to what he considers to be a particular problem of seventeenth-century science rather than an inference (p. 48 n. 3). For the historiography of the apparent tension between Boyle's ontology and his epistemology, see Rose-Mary Sargent, *Diffident Naturalist*, pp. 2–3. For the failure of the new philosophy to justify adequately appeals to insensible objects, see Alan Chalmers, "The Lack of Excellency of Boyle's Mechanical Philosophy," *Studies in History of the Philosophy of Science* 24 (1993), pp. 541–564; Christoph Meinel, "Early Seventeenth-Century Atomism: Theory, Epistemology, and the Insufficiency of Experiment," *Isis* 79 (1988), pp. 68–103.

34. Boyle, *Sceptical Chymist*, *Works*, vol. 1, p. 516.

siderer. And this is strongly confirmed by experience, which witness-
eth, that, in almost all ages and countries, the generality of philosophers, and contemplative men, were persuaded of the existence of a
Deity.[35]

I should point out here that the insensible objects of nature and God
do not have the same epistemological status for Boyle. As a Christian,
Boyle believed that revelation assured him of God's existence, whereas
he had no revelation about the existence of corpuscles (or any other
imperceptible material objects); while revelation confirmed his belief that
God exists, his claims about corpuscles remained hypothetical. My point
is that in both cases, even though the cause of an effect might not be
perceptible to sensory perception, the effect is. And just as a "heedful
and intelligent contemplator" of nature is led from a consideration of
nature to a consideration of the cause (or Creator) of nature,[36] so, too,
is the natural philosopher led from an observed effect to a consideration
of the possible causes of that effect. Further, Boyle made it clear that he
was aware that there is a connection between the insensible things of
this world and those of the next when he noted that the Epicureans
might argue that they have reason to affirm the existence of matter and
deny the existence of God because the former is confirmed by their
senses. But, he answered, "an Atome is as well invisible as the Deity."[37]
Visibility to the human eye, Boyle argued, is not necessary to the existence of an atom, a corpuscle of air, or of the effluviums of a loadstone.[38]
Nor is being perceptible by the senses necessary to the existence of God.

Boyle believed that the fact that the particular entities that cause
certain effects are imperceptible to the senses should not keep the natural
philosopher from attempting to learn as much as possible about the
nature of those causes by investigating the conditions under which the
effects are brought about and other conditions in which the effects are
not brought about.[39] The natural philosopher, then, should seek plausible explications of the phenomena even though those explications involve speculations about entities not perceptible to the senses. In the

35. Boyle, *Christian Virtuoso, Works*, vol. 5, pp. 515–516. All of Boyle's
    Disquisition about the final Causes of natural Things, of course, was
    devoted to the same point.
36. Boyle, *Christian Virtuoso, Works*, vol. 5, p. 516.
37. Boyle Papers, vol. 2, fol. 12.
38. Boyle, *Advices, Works*, vol. 4, p. 450.
39. See, for example, Boyle, *Essays of the Strange Subtilty, Works*, vol. 3, pp.
    659–732.

next chapter, I show how in Boyle's view the limits of human understanding make it impossible for the natural philosopher to judge exactly how insensible entities produce effects. Here, my purpose has been to show that in Boyle's view, the fact that the entities are insensible does not keep the natural philosopher from investigating them by studying "what changes are made in the patient, to bring it to exhibit the phænomena."[40]

40. Boyle, *Excellency and Grounds of the Mechanical Hypothesis, Works,* vol. 4, p. 73.

7

The Limits of Reason and Knowledge of Nature

I have argued that Boyle's primary motivation for writing *Discourse of Things above Reason* was a theological one, and the work was in fact considered to be a theological work at the time. It was listed as a theological work in catalogues issued by his publishers in his later years, as well as classified as theological by subsequent compilers of collections of his works.[1] Nevertheless, Boyle explicitly presented the work as a *philosophical* essay, not a theological one, claiming that the category of things transcending reason is not limited to revealed truths but extends to matters of natural philosophy as well.

Boyle's position here was certainly not original; as I pointed out in chapter 1, it can be traced back at least as far as Irenaeus (born c. 126) who, arguing in a theological context not unlike that which prompted Boyle's *Things above Reason,* supported his claim that certain Christian doctrines are genuinely mysterious by pointing out that some natural phenomena are also mysterious.[2] As Étienne Gilson has put it, this "insistence on the deficiency of natural knowledge introduced, if not for the first time, at least with a force that was then new, what will remain, up to the time of Montaigne's *Apology for Raymond Sebond,* the favorite theme of a certain type of Christian apologetics."[3] The deficiency of

---

1. See, for example, Richard Boulton, *The Theological Works of the Honourable Robert Boyle, Esq.; epitomiz'd* (London, 1715). I am grateful to an anonymous reader for having brought this general point to my attention.
2. Irenaeus, *Against Heresies,* translated by Alexander Roberts and James Donaldson, in *The Ante-Nicene Fathers: Translations of the Writings of the Fathers down to A. D. 325,* edited by Alexander Roberts and James Donaldson, revised by A. Cleveland Cox. 10 vols. American Reprint of the Edinburgh Edition (Grand Rapids, MI: Wm. B. Eerdmans, 1956), vol. 1, p. 399.
3. Étienne Gilson, *History of Christian Philosophy in the Middle Ages* (New York: Random House, 1955), p. 22. Only fifteen years after Boyle's *Things above Reason,* John Toland, the deist, turned the argument on its head,

natural knowledge was certainly a favorite theme of Boyle's a hundred years after the publication of Montaigne's *Apology* in 1568. In the hands of one of the founders and foremost members of the Royal Society of England, it took on a new significance, and received an application that helped define the limits of the new natural philosophy.

Tracing Boyle's views on the limits of reason in the context of natural philosophy and the application of those views to particular questions will require an appeal to passages in a variety of Boyle's works. The best place to begin is the now-familiar *Discourse of Things above Reason* and works related to this particular theme, and an investigation of Boyle's categories of the incomprehensible, the inexplicable, and the unsociable as they pertain to the knowledge of nature. In the following discussion, it will become obvious that Boyle often conflated discussions of questions about natural philosophy with theological considerations; this tendency only emphasizes the point that for him, the two were inseparable.

The Incomprehensible, the Inexplicable, and the Unsociable

Just as Boyle claimed that in theology, some matters surpass human understanding because they are incomprehensible, inexplicable, or unsociable (contradictory), so, too, he claimed that in natural philosophy, some matters surpass human understanding for the same reasons. It is important to emphasize that in natural philosophy, as in theology, these categories are not mutually exclusive; in the discussion that follows it will become obvious that some objects of knowledge fall into two or even all three categories. Nevertheless, each category has a significance of its own that justifies following Boyle in treating them separately.

---

arguing that nothing in Christianity is any more mysterious than many everyday natural phenomena in *Christianity not Mysterious; or, A Treatise shewing that there is nothing in the Gospel contrary to reason, nor above it, and that no Christian doctrine can be properly call'd a mystery* (London, 1696). See Robert E. Sullivan, *John Toland and the Deist Controversy* (Cambridge: Harvard University Press, 1982); Stephen H. Daniel, *John Toland: His Methods, Manners, and Mind* (Kingston and Montreal: McGill-Queen's University Press, 1984); and J.A.I. Champion, *The Pillars of Priestcraft Shaken: The Church of England and Its Enemies, 1660–1730* (Cambridge: Cambridge University Press, 1991).

The Incomprehensible

Boyle's most frequent examples of questions involving natural knowledge that are incomprehensible were those that involved the concept of infinity, a fact that is not surprising. As I have already shown, his emphasis in *Things above Reason* was on the disparity between God's infinite and man's finite understanding. In fact, it was in the contemplation of infinites that Boyle most clearly perceived the limits of human intellects. One of the interlocutors, Eugenius, expressed the frustration he felt when he contemplated things involving the concept of infinity by remarking that

> I have not the vanity to think, that the weakness of my reason ought to make another diffident of the strength of his: but as to myself, ... having exerted the small abilities I had, to clear up to myself some of the difficulties about infinites, I perceived, to my trouble, that my speculations satisfied me of nothing so much, as the disproportionateness of those abstruse subjects to my reason.[4]

In this passage, Eugenius acknowledged that the difficulty might be the result of the weakness of his own reason rather than the weakness of human understanding in general. Elsewhere, however, Boyle claimed that questions about infinity posed insurmountable problems for even such "excellent persons" as Galileo and Descartes, noting that the "indivisible and infinite are things, that do so swallow up the mind of man, that he scarce knows what to pitch on, when he contemplates them."[5]

He made much the same point in the *Appendix* to *Christian Virtuoso,* where he noted that

> because there belongs to numbers a kind of infinity, or somewhat that is near of kin to it we must, if we want not attention or skill, discover by the incompetency of our utmost endeavours the boundedness of our human understandings; since when we have proposed a determinate number as great as we are able to ascend to, yet we are sensible that it is not absolutely the greatest of numbers, since it may be augmented by the addition of an unite, or an aggregate of unites.[6]

And in the *Christian Virtuoso* itself, he noted that many atheists accuse those who assert God's existence as asserting a doctrine that is "encumbered with inextricable difficulties." But, Boyle insisted in that work,

4. Boyle, *Things above Reason, Works,* vol. 4, p. 412.
5. Boyle, *Reason and Religion, Works,* vol. 4, pp. 160–161, quote, p. 160, where Boyle was referring to an anonymous French mathematician who had discussed Galileo's proof that a point is equal to a circle.
6. Boyle, *Appendix* to *Christian Virtuoso, Works,* vol. 6, p. 696.

most of the perplexing difficulties, the atheists lay so much stress on, do not proceed from any absurdity contained in the tenet of the theists, but from the nature of things; that is, partly from the dimness and other imperfections of our human understandings, and partly from the abstruse nature, that, to such bounded intellects, all objects must appear to have, in whose conception infinity is involved, whether that object be God, or atoms, or duration, or some other thing that is uncausable. For, however we may flatter ourselves, I fear we shall find, upon strict and impartial trial, that finite understandings are not able clearly to resolve such difficulties, as exact a clear comprehension of what is really infinite.[7]

From the philosophical point of view, Boyle's emphasis on the difficulties that infinites pose for human comprehension is significant. It indicates that he did not believe that it is possible for human beings to understand fully the nature of space and time, and indeed he suggested that space might be one of those "privileged" matters that are exempt from the ordinary rules of reasoning.[8] Similarly, he thought the human mind incapable of grasping the nature of time, twice citing St. Augustine on the subject.[9] From the point of view of Boyle's conception of the task of the seventeenth-century natural philosopher, however, the category of those things that Boyle considered to be inexplicable was of more significance than those things specifically requiring the contemplation of infinites.

The Inexplicable

The category of the inexplicable consisted of those things whose causes were unknown. For Boyle, a phenomenon was inexplicable if human beings were unable to provide even a plausible explanation of it; inexplicable things were those things concerning which "the *modus operandi* is

7. Boyle, *Christian Virtuoso, Works,* vol. 5, p. 515.
8. Boyle, *Things above Reason, Works,* vol. 4, pp. 459–460. In this regard, Boyle cited Gassendi's view that space is neither accident nor substance; for Gassendi's views, see Margaret J. Osler, *Divine Will and the Mechanical Philosophy: Gassendi and Descartes on Contingency and Necessity in the Created World* (Cambridge: Cambridge University Press, 1994), pp. 182–183. For Gassendi's influence on Boyle, see Osler, "The Intellectual Sources of Robert Boyle's Philosophy of Nature: Gassendi's Voluntarism and Boyle's Physico-Theological Project," in *Philosophy, Religion, and Science, 1640–1700,* edited by Richard Kroll, Richard Ashcraft, and Perez Zagorin (Cambridge: Cambridge University Press, 1992), pp. 178–198.
9. Boyle, *Reason and Religion, Works,* vol. 4, p. 173; *Theological Distinction, Works,* vol. 5, p. 544.

beyond our comprehension," and he meant "not . . . only the true and certain *modus operandi,* but even an intelligible one."[10] The significance of this distinction between an intelligible and a true explanation will become clear when I discuss his views on the status and role of hypotheses, showing that Boyle's goal in natural philosophy was to provide an intelligible explication of phenomena whenever possible, and that he rarely went on to claim that his explication was true and that alternative explanations were false.

In any event, some phenomena resisted even a plausible explication. As Boyle put it,

> when by attentively considering the attributes and operations of things, we sometimes find, that a thing hath some property belonging to it, or doth perform somewhat, which by reflecting on the beings and ways of working that we know already, we cannot discern to be reducible to them or derivable from them, we then conclude this property or this operation to be inexplicable; that is, such, as that it cannot so much as in a general way be intelligibly accounted for.[11]

The category of the inexplicable ranged from ordinary everyday objects to the highest of secondary causes. Among Boyle's examples we find "the manner of operating, whereby several bodies perform what we well know they bring to pass," how the cohesion of the smallest particles of matter is effected, how the human soul can move the human body, how understanding and will act upon one another, and how the human memory operates.[12] "As much as we pride ourselves in our understanding," he noted, "we may daily observe that the actions and performances of a senseless grain of corn for instance, or of a kernel of an apple, surpass our comprehension."[13] Other examples of ordinary phenomena for which no intelligible explanation had yet been found included poisons and their antidotes, the peculiar properties of quicksilver, and the strange effects fancies and cravings of pregnant women that had been observed to have on the fetus.[14] And in an unpublished essay entitled

---

10. Boyle, *Things above Reason, Works,* vol. 4, p. 416.
11. Boyle, *Things above Reason, Works,* vol. 4, p. 422.
12. Boyle, *Advices, Works,* vol. 4, p. 454. Boyle was fascinated by memory; see, for example, John Beale's letters to Boyle, 4 Oct. 1661 (*Works,* vol. 6, pp. 325–326), and 29 Sept. 1663 with additions made 2 Oct. 1663 (*Works,* vol. 6, pp. 330–342).
13. Boyle, *Appendix* to *Christian Virtuoso, Works,* vol. 6, p. 683.
14. Boyle, "Humani intellectus imperfectio nativo suo lumine detecta," Boyle Papers, vol. 16, fols. 72–85. Here, Boyle considered a phenomenon to be unintelligible if no mechanical explanation could be found for it. I am

"An Essay of Various Degrees or Kinds of the knowledge of natural things," he claimed that although the human intellect is capable of ascertaining some of "the lower degrees or stops in the scale of causes," human beings should "forbear aspiring to the knowledge of the highest among second causes."[15]

It was in his discussion of inexplicable things that Boyle most often and most explicitly remarked on the parallel between theological matters and matters of natural philosophy. He noted, for example, that

> it is both strange and unjust, that our quarrellers with religion should require, that we, that embrace it, should give them clear and direct solutions of all the difficulties, which . . . they are able to object against theological mysteries, whilst in the mean time themselves are not able to give a satisfactory account of the difficulties, that attend the distinct explication of merely corporeal, and perhaps too very familiar subjects. Insomuch, that those very men rigorously exact that we should explain to them the mysteries of the Trinity, or the incarnation of the second person of it, that cannot clearly shew, what keeps the parts of the least portion of matter together, or how, when they have a mind to speak, their own souls can direct the animal spirits to all the little organs, that are necessary to the formation of a vocal discourse. . . .
>
> We must not suppose that, at least in our present state, our reason and other faculties are given us, to reach all that are knowable, even as to corporeal creatures; but only things that are in such a sphere of intelligibility, that they are proportioned to our present faculties, and convenient for our notice in our present state and condition.[16]

And in another passage he argued that

> if in things corporeal, that are the familiar objects of our senses, we are often reduced to confess our ignorance of the modes of their existing or operating, I hope it will not be denied, that to a being wholly unapproachable to our senses, natural theology may be allowed to ascribe some things, whose modus is not attainable by our understanding: as the divine prescience of future contingents, which

---

grateful to Dr. Michael Hunter for having sent me a translation of this manuscript, which is to be included along with other previously unpublished Boyle manuscripts in the forthcoming Pickering edition of Boyle's *Works*, edited by Michael Hunter and Edward B. Davis. The inclusion of these manuscripts, a project supervised by Hunter, has been funded by the Leverhulme Trust.

15. Boyle Papers, vol. 8, fol. 167. I discuss this essay in more detail when I discuss the limits of mechanism later in this chapter.
16. Boyle, *Appendix* to *Christian Virtuoso*, *Works*, vol. 6, pp. 695–696.

as it were impious to deny, as to the truth of the thing, so I fear it is impossible to explicate, as to the modus of it.[17]

In the previous chapter, I noted that Boyle believed that it is legitimate to infer the existence of insensibles on the grounds of observable effects. Believing that a particular entity exists, however, is not the same as explaining how it produces the effects it does; human beings do not understand how a loadstone operates or how it is that God brings about certain effects in the world. "There is no necessity," he commented, "that intelligibility to a human understanding should be necessary to the truth or existence of a thing."[18] Nevertheless, if there is an observable effect, a cause must be assumed, even if the exact nature of that cause is not known:

> If there be an effect, that we discern must proceed from such a cause, or agent, we may conclude that such a cause there is, though we do not particularly conceive, how or by what operation it is able to produce the acknowledged effect.[19]

Where knowledge of the natural world was concerned, there were, in Boyle's view, things that could not be understood fully because they are incomprehensible (such as space and time) and things that could not be understood fully because they are inexplicable (such as how the minute particles of matter adhere and how the human soul can move the human body). Where these two categories were concerned, Boyle made no significant distinction between their applications in theology and in natural philosophy. His third category of theological things above reason, however, the category of unsociable or contradictory truths, posed a distinct difficulty where natural philosophy was concerned.

The Unsociable

It would seem that there should be no unsociable truths in natural philosophy because Boyle's primary requirement for admitting contradictory truths into the body of knowledge was that at least one of those truths must be revealed, and (as I showed in chapter 6), except for some few truths concerning the origin and ultimate fate of the world, Boyle rejected scriptural revelation as a way in which knowledge of the natural world could be attained. Indeed it is the case that although Boyle argued that some revealed truths may violate the law of noncontradiction,

17. Boyle, *Advices, Works,* vol. 4, p. 454.
18. Boyle, *Appendix* to *Christian Virtuoso, Works,* vol. 6, p. 694.
19. Boyle, *Advices, Works,* vol. 4, p. 455.

he explicitly refused to admit the possibility of contradictions where knowledge of the natural world was concerned. Boyle made this clear when he considered an interlocutor's objection to the admitting of contradictory propositions into our body of knowledge. Posing the objection, Pyrocles asked,

> can any thing seem more unreasonable, than to embrace opinions, that contradict the rules of reason? which practice, if it be once allowed, why should we trouble ourselves to investigate what is congruous or incongruous to reason, since the making a discovery, that an opinion is repugnant to it, will not assure us of that opinions being false?[20]

Boyle, in the person of Arnobius, responded that where the matters being considered did not involve revelation but were in the jurisdiction of ordinary reason, he would in "no way disallow the rejecting of opinions, that are found contrary to those rules of reason, at the framing of which the things opined about were duly taken into consideration."[21] His point was that his argument for exempting some revealed truths from the law of noncontradiction was grounded (at least partially) in his belief that the rules of reasoning employed to judge of ordinary matters should not be applied to many revealed truths, and that to judge a revelation by the rules of reasoning would be to extend those rules beyond their proper boundaries. In fact, where knowledge obtained by natural means was concerned, Boyle positively required "that the propositions framed about them be estimated by the common dictates of reason."[22]

There were, then, in Boyle's view, no *genuinely* "unsociable" truths in natural philosophy, as there were, at least as judged by finite human understanding, in matters involving revealed truths. Nevertheless, he claimed that in natural philosophy there are contradictory propositions that pose insurmountable problems for human beings because of the limits of human understanding. The difference is that in theology, both contradictory propositions must be accepted as true because of the authority that lies behind the revealed truth, whereas in natural philosophy, at least one of the propositions must, in theory, be rejected. The problem, as Boyle saw it, is that given the limits of human understanding, it is not always apparent which of the two propositions of natural philosophy should be rejected because both propositions are subject to unanswerable objections. Indeed, Boyle considered unanswerable objec-

---

20. Boyle, *Things above Reason, Works,* vol. 4, p. 457.
21. Boyle, *Things above Reason, Works,* vol. 4, p. 458.
22. Boyle, *Advices, Works,* vol. 4, p. 466.

tions to be the criterion of "unsociable propositions," and used the controversy about the endless divisibility of a straight line as an example:

> Since it is manifest, that a line of three foot, for instance, is thrice as long as a line of one foot, so that the shorter line is but the third part of the longer, it would follow, that a part of a line may contain as many parts as a whole, since each of them is divisible into infinite parts; which seems repugnant to common sense, and to contradict one of those common notions in *Euclid*, whereupon geometry itself is built. Upon which account I have ventured to call this third sort of things above reason asymmetrical or unsociable, of which eminent instances are afforded us by those controversies (such as that of the *compositio continui*) wherein which side soever of the question you take, you will be unable directly and truly to answer the objections, that may be urged to shew, that you contradict some primitive or some other acknowledged truth.[23]

In short, there may be two propositions of natural knowledge, both of which entail consequences that violate the law of noncontradiction, and the fact that in theory at least one of the propositions must be false is of no help. In such cases, Boyle said,

> we may be reduced either to reject inferences legitimately drawn from manifest or granted truths, or to admit conclusions, that appear absurd; if we will have all the common rules, whereby we judge of other things, to be applicable to infinites.[24]

Still, at least one of the propositions *must* be false. Boyle made this clear in *Reflections upon a Theological Distinction,* where he offered as an example the controversy concerning the endless divisibility of quantity. That matter is endlessly divisible was affirmed (and indeed proven) by Aristotle and "by several excellent geometricians besides." Nevertheless, many eminent mathematicians, and an even greater number of natural philosophers (most particularly the Epicureans and other atomists), "stiffly maintain the negative."

> In reality [Boyle continued] the assertions of these two contending parties are truly contradictory; since of necessity a straight line must be, at least mentally, divisible into parts that are themselves still further divisible; or it must not be so, and the subdivisions must at length come to a stop. And therefore one of the opposite opinions must be true. And it is plain to those, that have, with competent skill and attention, impartially examined this controversy, that the side, that is pitched upon, which soever it be, is liable to be exposed to such difficulties, and other objections, as are not clearly answer-

---

23. Boyle, *Things above Reason, Works,* vol. 4, p. 423.
24. Boyle, *Things above Reason, Works,* vol. 4, p. 409.

able, but confound and oppress the reason of those, that strive to defend it.[25]

Boyle did not limit his discussions of apparently contradictory propositions encountered in knowledge of the natural world to examples of problems involving infinites. He also observed that in natural philosophy,

> many things are judged to be contrary to the true laws or dictates of right reason; because men do either presume something to be manifestly true, that is not so, or do not know, or perhaps so much as suspect, the differing ways, by which a thing may be produced, or brought to pass, or may have a right to the title of true.[26]

The relationship of these "unsociable" truths in natural philosophy to theological controversies should be noted. The differing explications of controversial passages of scripture offered by two quarreling sects of Christians (for example, the Calvinists and the Arminians) might *each* imply contradictions. In fact, this is exactly the charge polemicists from each of these two groups leveled against the other, with Calvinists charging the Arminians of making God's decrees contingent on human actions (which contradicts God's omnipotence) and Arminians claiming that if the Calvinist interpretation were true, God would have created some souls simply in order to damn them (which contradicts God's goodness). Hence both interpretations would be vulnerable "to such difficulties, and other objections, as are not clearly answerable, but confound and oppress the reason of those, that strive to defend it." Nevertheless, in some cases (and this was one of them), *experience* could provide a criterion against which the different claims could be judged; Boyle himself rejected, at least tentatively, the theological determinism implied by Calvinism, and he did so on the grounds of the experience human beings have of the freedom of the will.[27] Similarly, in natural philosophy, Boyle presented experience as a criterion by which, in some cases, a judgment between two contradictory propositions, each based on reason alone, could be made. If experience presented evidence contrary to a judgment based solely on reason, the proposition based on

25. Boyle, *Reflections upon a Theological Distinction, Works,* vol. 5, p. 547. In Boyle's view, matter was infinitely divisible mentally, but matter was not infinitely divisible by natural agents physically, although God in his omnipotence might be able to divide physical matter infinitely; see *Origin of Forms and Qualities, Works,* vol. 3, p. 29.
26. Boyle, *Appendix to Christian Virtuoso, Works,* vol. 6, p. 711.
27. See, for example, Boyle, *Things above Reason, Works,* vol. 4, p. 409; *Advices, Works,* vol. 4, p. 466.

experience should be judged true and the assumption of reason judged false. As he put it,

> we ought to believe divers things upon the information of experience, (whether immediate, or vicarious) which, without that information, we should judge unfit to be believed, or antecedently to it did actually judge contrary to reason.[28]

As examples, he offered cases in which the effect seems disproportionate to the cause, such as the power of "a light black powder" to "throw down stone walls, and blow up whole castles," the power of "two or three grains of opium" to "stupify a whole body," and the venomous effects of the small amount of poison transmitted by the bite of a "mad dog."[29] Other examples included cases where previous philosophers, uninformed by experience, had judged falsely, as when the Aristotelians judged the "torrid zone" to be uninhabitable[30] and when Aristotle judged that the heavenly bodies are not subject to generations and corruptions.[31] In such cases, the conclusions of reason must give way to the evidence of experience. It is important to note, however, that such instances reflected the acquisition of a new matter of fact. They reflected situations in which Boyle believed human reason capable of reaching a correct judgment. In the remainder of this chapter I examine the influence of his views on the limits of human understanding when the natural philosopher judges those things that are *not* matters of fact, and show how this affected his goals in natural philosophy.

The Task of the Natural Philosopher

Clearly, Boyle believed that there are mysteries in the realm of knowledge of the natural world, just as there are mysteries in the spiritual realm, and his views concerning the proper goals of the new natural philosophy were influenced by this belief. Although the phenomena themselves might be incomprehensible, inexplicable, or even at times apparently contradictory, the goal of the natural philosopher, as Boyle saw it, was to provide intelligible and consistent explications of the phenomena of nature. In other words, Boyle thought that it was the task

28. Boyle, *Christian Virtuoso, Works,* vol. 5, p. 526.
29. Boyle, *Christian Virtuoso, Works,* vol. 5, p. 527.
30. Boyle, *Reason and Religion, Works,* vol. 4, p. 181.
31. Boyle, *Christian Virtuoso, Works,* vol. 5, p. 528.

of the natural philosopher to devise hypotheses by means of which the phenomena could be understood in terms of natural processes.

Explanations Must Be Natural

Before describing Boyle's views on the formation of and evaluation of hypotheses, it is worth pausing for a moment to emphasize his insistence that explanations of the phenomena of nature must themselves be *natural* explanations. Although this may appear obvious and almost trivial to the twentieth-century reader, it was by no means taken for granted in Boyle's day.[32] In denying that the apparently empty space in the Toricellian experiment was really empty, for example, the Aristotelian Franciscus Linus argued that the space was really filled with a rarefied (and invisible) corporeal substance (a "funiculus"), most likely mercury. In his explication of this funiculus, Linus claimed that the substance was "virtually extended," and argued that, incomprehensible as "virtual extension" may be, it is not impossible when God's omnipotence is considered. To this, Boyle replied that although no one had more reverence for divine omnipotence than he himself did,

> our controversy is not [about] what God can do, but about what can be done by natural agents, not elevated above the sphere of nature. . . . [Our hypothesis of the spring and weight of the air] would need no other advantage, to make it be preferred before our adversaries, than that in ours things are explicated by the ordinary course of nature, whereas in the other recourse must be had to miracles.[33]

Boyle made the same point in his controversy with Hobbes. Hobbes had objected to Boyle's *History of Fluidity and Firmness,* arguing that the small size of the constituents of some bodies could not account for fluidity, and basing his objection on the grounds that matter is infinitely divisible, at least by divine omnipotence. Boyle responded that in the first place, Hobbes had misinterpreted his explanation, which had rested on the motion of the particles, not their size. And in the second place,

---

32. For a smorgasbord of non-natural explanations prevalent in the culture at large, see Keith Thomas, *Religion and the Decline of Magic: Studies in Popular Beliefs in Sixteenth and Seventeenth Century England* (London: Weidenfeld and Nicolson, 1971). Specific examples of non-natural explanations offered by Boyle's fellow philosophers are offered below.

33. Boyle, *Defence of the Doctrine, Works,* vol. 1, p. 149. For background information on Line, see chapter 6, note 2.

when Mr. *Hobbes* has recourse to what God can do (whose omnipotence we have both great reason to acknowledge) it imports not
to the controversy about fluidity to determine what the almighty
Creator can do, but what he has actually done.[34]

Boyle was certainly correct in pointing out to Linus and Hobbes that
no one had more respect for God's omnipotence than he himself did.
Indeed, Boyle's emphasis on God's will and power is of such significance
to the thesis of this book that it warrants a separate treatment in the
next chapter. For the moment, however, it is enough to note that I see
no conflict between Boyle's belief that God can (and may) intervene in
the course of nature and his belief that the explanations of a natural
philosopher should be limited to natural ones. Regardless of whatever
way or ways God might cause the phenomena, the natural philosopher's
explications must be well grounded in the phenomena, and how God
bring might about particular effects (whether in the normal course of
nature or by his supernatural intervention in that course) is not manifest
in the phenomena.[35]

Explanations Must Be Grounded in the Phenomena

A good explanation must be not only a natural one, but also one well-
grounded in the phenomenon to be explained. Boyle believed that the
formation of premature or purely speculative hypotheses was one of the
most serious mistakes a natural philosopher could make. Indeed, in one
passage he went so far as to say that

> it was not my chief design to establish theories and principles, but
> to devise experiments, and to enrich the history of nature with
> observations faithfully made and delivered; that by these and the
> like contributions made by others, men may in time be furnished
> with a sufficient stock of experiments, to ground hypotheses and
> theories on.[36]

---

34. Boyle, *An Examen of Mr. Hobbes's "Dialogus Physicus de Naturâ Aëris,"*
    Works, vol. 1, p. 236.
35. Boyle was not as consistent in his refusal to appeal to divine omnipotence
    as one would like. In *Final Causes* (*Works*, vol. 5, p. 514) he noted that
    although it is not easy to conceive how God can, "by so simple an instru-
    ment as local motion, be able to direct a multitude of agents, as numerous
    as the bodies that make up a world, to act as regularly, as if each of them
    acted upon its own particular design, and yet all of them conspired to obey
    the laws of nature. But if we consider, that it is to God, that is an
    omniscient and almighty Agent, that this great work is ascribed, we shall
    not think it incredible."
36. Boyle, *Defence of the Doctrine, Works*, vol. 1, p. 121.

Similarly, he warned against premature or purely speculative hypotheses in *Considerations touching Experimental Essays:*

> If men could be persuaded to mind more the advancement of natural
> philosophy than that of their own reputations, it were not, me-
> thinks, very uneasy to make them sensible, that one of the consider-
> ablest services, that they could do mankind, were to set themselves
> diligently and industriously to make experiments and collect obser-
> vations, without being over-forward to establish principles and
> axioms, believing it uneasy to erect such theories, as are capable to
> explicate all the phænomena of nature, before they have been able
> to take notice of the tenth part of those phænomena, that are to be
> explicated.[37]

The Baconian tone of such statements is unmistakable. Nevertheless, Boyle should not be misunderstood as a "naive Baconian" (and for that matter, neither should Bacon).[38] Although the gathering of facts, and especially facts based on experiments, was foundational for both Boyle and Bacon, the purpose of such fact-gathering was to enable the natural philosopher to formulate hypotheses that were well-grounded in the phenomena. Although the hypotheses thus formed might be provisional and subject to revision or even subsequent rejection, they were an important aid to the natural philosopher. Boyle continued with the statement that he did not at all

> disallow the use of reasoning upon experiments, or the endeavour-
> ing to discern as early as we can the confederations, and differences,
> and tendencies of things. . . . In physiology it is sometimes condu-
> cive to the discovery of truth, to permit the understanding to make
> an hypothesis, in order to the explication of this or that difficulty,
> that by examining how far the phænomena are, or are not, capable
> of being solved by that hypothesis, the understanding may, even by
> its own errors, be instructed. . . . That then, that I wish for, as to
> systems, is this, that men, in the first place, would forbear to estab-
> lish any theory, till they have consulted with (though not a fully
> competent number of experiments, such as may afford them all the
> phænomena to be explicated by that theory, yet) a considerable
> number of experiments, in proportion to the comprehensiveness of

---

37. Boyle, *Experimental Essays, Works,* vol. 1, pp. 302–303.

38. Margery Purver places undue emphasis on the fact-gathering aspect of the "new philosophy" in *The Royal Society: Concept and Creation* (Cam-bridge: M.I.T. Press, 1967). For more accurate portrayals of Baconianism within the Royal Society, see Michael Hunter and Paul B. Wood, "Towards Solomon's House: Rival Strategies for Reforming the Early Royal Society," *History of Science* 24 (1986), pp. 49–108. See also Marie Boas [Hall], "Boyle as a Theoretical Scientist," *Isis* 41 (1950), pp. 261–268.

the theory to be erected on them. And, in the next place, I would have such kind of superstructures looked upon only as temporary ones; which though they may be preferred before any others, as being the least imperfect, or, if you please, the best in their kind that we yet have, yet are they not entirely to be acquiesced in, as absolutely perfect, or uncapable of improving alterations.[39]

For Boyle, matters of fact, once established, were true, whereas a hypothesis was simply a possible explanation of the phenomena, an exploration of the way or ways in which the phenomena might be effected in the ordinary course of nature. He has been criticized, most recently by Steven Shapin and Simon Schaffer, for not giving "the criteria and rules for establishing hypotheses."[40] Such criticism seems unwarranted in view of the passages I have just quoted (as well as in view of his explicit discussion of what constitutes a good or an excellent hypothesis, which I shall discuss shortly). Shapin and Schaffer have also missed the point, I believe, in their related criticism that Boyle inconsistently treated the spring and pressure of the air sometimes as a hypothesis and at other times as an established fact.[41] Boyle viewed the identification of causes to be an activity that involved steps from lower-level explanations to more general ones, a process by which a lower-level hypothesis might become, as a result of carefully contrived experiments, a matter of fact.[42]

In the passages I have quoted here, Boyle criticized "system-builders" not just for devising theories that were premature by being more comprehensive than warranted by the experimental data; he also criticized them for considering those theories to be "perfect, or uncapable of improving alterations." In Boyle's view, the natural philosopher must always be prepared to modify or abandon a hypothesis in the event that future matters of fact that indicate a weakness in or the incompleteness of the hypothesis are discovered. In *Origin of Forms and Qualities*, for example, Boyle noted that "it becomes a naturalist, not only to advise hypotheses and experiments, but to examine and improve those that are already found out." In this work, Boyle introduced the corpuscular explanation

---

39. Boyle, *Experimental Essays, Works*, vol. 1, p. 303.
40. Steven Shapin and Simon Schaffer, *Leviathan and the Air-Pump: Hobbes, Boyle, and the Experimental Life* (Princeton: Princeton University Press, 1985), p. 51.
41. Shapin and Schaffer, *Leviathan and the Air-Pump*, p. 50.
42. For a more sophisticated understanding of Boyle and "matters of fact" than that of Shapin and Schaffer as expressed in *Leviathan and the Air-Pump*, see Rose-Mary Sargent, *The Diffident Naturalist: Robert Boyle and the Philosophy of Experiment* (Chicago: University of Chicago Press, 1995), esp. pp. 132–134.

of forms and qualities as a hypothesis "to be confirmed or disproved" by experiments.[43] Even in his defense of his own hypothesis of the spring and weight of the air, a hypothesis relatively low on any scale of comprehensiveness insofar as it extended no further than to some particular phenomena involving air and one which he considered to be well-established, Boyle noted that some of the experimental results were not exactly those that had been expected. And although the variations were slight, and were, Boyle thought, most likely due to the inexactness of the experiments, he withheld any conclusion as to the universality of the hypothesis until such a time as future experiments might justify such a claim.[44] That Boyle was indeed willing, at least in theory, to modify his own hypotheses is further indicated by his statement that if anyone should raise objections that he considered valid to any of his "opinions or explications," he would "alter, mend, supply, vindicate or retract" those opinions or explications.[45] A number of scholars have described the trend away from dogmatic claims of certain knowledge toward a more modest probabilism and fallibilism in the seventeenth century. Boyle's views on hypotheses are a prime example of this trend.[46]

Good and Excellent Hypotheses

Boyle knew that in many cases, several alternative hypotheses could account for various phenomena of nature, and that the proponents of these various hypotheses believed each to be well-grounded in the phenomena. He had explicit and clear-cut criteria for preferring one of the hypotheses (the corpuscular hypothesis) and rejecting the others (such as the substantial forms of the Schoolmen and the hylarchic principle of the Cambridge Platonists). These criteria are listed in Boyle's manuscript notes on "Good and Excellent Hypotheses,"[47] and I summa-

---

43. Boyle, *Origin of Forms and Qualities*, *Works*, vol. 3, pp. 3, 14.
44. Boyle, *Defence of the Doctrine*, *Works*, vol. 1, p. 159.
45. Boyle, *Defense of the Doctrine*, *Works*, vol. 1, p. 122.
46. See, for example, Henry G. van Leeuwen, *The Problem of Certainty in English Thought, 1630–1690* (The Hague: Martinus Nijhoff, 1963), especially Richard H. Popkin's introduction to this work; Barbara Shapiro, *Probability and Certainty in Seventeenth-Century England* (Princeton: Princeton University Press, 1983), and Ian Hacking, *The Emergence of Probability: A Philosophical Study of Early Ideas about Probability, Induction and Statistical Inference* (Cambridge: Cambridge University Press, 1975).
47. Boyle Papers, vol. 35, fol. 202. This manuscript has been published by

rize them here: A *good* hypothesis (or explication) must be intelligible. It must contain nothing impossible or manifestly false. It must not suppose anything that is unintelligible, impossible, or absurd. It must be self-consistent. It must be sufficient to explain the phenomena being investigated, and it must be consistent with related phenomena and not contradict any other known phenomena. An *excellent* hypothesis must be all of the above, and in addition it must be based on sufficient evidence, it must be the simplest of all the good hypotheses that can be formulated to explain the phenomena (or at least contain nothing superfluous), it must be either the only hypothesis that can account for the phenomena or the hypothesis that best accounts for the phenomena, and it must allow the natural philosopher to predict future phenomena by which the hypothesis may be confirmed or disconfirmed.

Boyle also expressed these criteria in a mnemonic version, which I include here for its intrinsic interest, as well as to serve as a succinct summary:

1. To frame a good Hypothesis, one must see First, *that* it clearly Intelligible be.

2. Next *that* it nought assume, nor do suppose That flatly dos any known Truth oppose.

3. Thirdly, that with itself it do consist So that no One part, th'other do resist.

4. Fourthly, Fit and sufficient it should be, T'*explain* all the *Phænomena* that we Upon good grounds, may unto It refer: Or those at least, that do the Chief appear.

5. Fifthly the Framer carefully must see That with the Rest, it do at least *agree*, And contradict no known Phænomena Of th' Universe, or any Natural Law.

6. Sixthly, An Hypothesis to be Excellent, Must not beg a præcarious Assent; But be built on Foundations Competent.

7. Next of all good, the Simplest it must be: At least from all that is superfluous, free.

8. Eighthly, It should the only be, that may The given *Phænomena* & so wel display.

9. Ninthly, It should inable us to foreshow The' Events that will, from welmade Tryals flow.[48]

M.A. Stewart, editor, in *Selected Philosophical Papers of Robert Boyle* (Indianapolis, IN: Hackett Publishing Co., 1991), p. 119.

48. Boyle Papers, vol. 36, fol. 57v. This version, as well as a third, may be found in Marie Boas Hall, *Robert Boyle on Natural Philosophy: An Essay with Selections from his Writings* (Bloomington: Indiana University Press, 1965), pp. 134–135.

Evaluation of Alternative Theories of Matter

It is important to stress three things about these criteria before proceeding to their specific applications. First is their secular nature. In his rejections of various hypotheses, Boyle often candidly noted that he had theological considerations in mind, and where they are relevant I shall note them. Nevertheless, Boyle did not rely on the theological reasons for rejecting a hypothesis. Instead, he usually explained that he would limit himself to objections appropriate to a natural philosopher rather than appeal to objections appropriate to a Christian, and the objections appropriate to a natural philosopher were almost always (at least implicitly) related to his criteria for "good and excellent" hypotheses.

The second thing that must be stressed before examining Boyle's evaluation of competing hypotheses – and this is the single most striking and instructive point to be gleaned from these criteria – is *the extent to which some of these criteria for a good explanation of the phenomena differ from Boyle's conception of the nature of some of the phenomena to be explained,* for these differences indicate the disproportion between what, in Boyle's view, is "knowable" (to God) and the knowledge of the phenomena that human understanding can hope to achieve. In other words, the best hypothesis might or *might not* accurately reflect reality; it might or might not be a *true* explanation.

The third thing to be stressed is that because of this probable (or at least possible) disparity between the phenomena themselves and the hypotheses formulated to explain them, Boyle did not consider himself to be in a position to claim that any of the more general hypotheses of nature was true, or that any was false. Only on rare occasions, when an extremely low-level explanation of some natural phenomenon was involved, did he venture to declare that his own explanation had achieved the status of a matter of fact considered to be true and that an alternative explanation was false, a point I shall discuss more fully below.[49]

Intelligibility

One of the criteria for a good hypothesis, Boyle claimed, is that it must be intelligible. It should be noted that the claim that a good explication of natural phenomena must be intelligible, even if in reality the phenom-

---

49. For a discussion of Boyle's view of the distinction between general and intermediate causes, see Sargent, *The Diffident Naturalist,* p. 97.

ena themselves are not caused in the way most intelligible to human understanding, directly parallels Boyle's view that a good explication in theology must be intelligible, even if the mystery being explained is itself incomprehensible. In the *Appendix* to *Christian Virtuoso,* Boyle stated that he did not think himself obliged

> to have the same regard and respect for the explications, that the schoolmen and many other divines give us of the mysteries of Christianity, that I have for the articles themselves; since for the mysteries I have the divine authority of the revealer, who can oblige my faith to assent even to dark truths: but for the expositions and consequences, I have but human authority: and though clearness is not always necessary to divine mysteries, yet it may be justly exacted in a good explication, since to explicate a thing, is to render it at least intelligible.[50]

Just as Boyle made a distinction between the "clearness" of a divine mystery and an explication of that mystery, he also distinguished in natural philosophy between the intelligiblity of the causes of natural phenomena and the intelligibility of plausible explications of those phenomena. The phenomena themselves may not in reality be caused in a way that is intelligible to human understanding, but the explanations that human understanding devises to aid and assist the natural philosopher in delving as deeply as possible into the phenomena must be intelligible. In *Usefulness of Experimental Natural Philosophy,* Boyle argued that whether God is the author of nature (as he himself believed), or whether natural phenomena are the result of chance, there was no reason to believe that phenomena are caused in the most intelligible way:

> There is no reason to imagine, that chance considered what manner of their production would be the most easily intelligible to us. And if God be allowed to be, as indeed he is, the author of the universe, how will it appear that he, whose knowledge infinitely transcends ours, and who may be supposed to operate according to the dictates of his own immense wisdom, should, in his creating of things, have respect to the measure and ease of human understandings; and not rather, if of any, of angelical intellects. So that whether it be to God or to chance that we ascribe the production of things, that way may often be fittest or likeliest for nature to work by, which is not easiest for us to understand.[51]

50. Boyle, *Appendix* to *Christian Virtuoso, Works,* vol. 6, p. 713.
51. Boyle, *Usefulness of Experimental Natural Philosophy, Works,* vol. 2, p. 46. He made the same point in *Appendix* to *Christian Virtuoso, Works,* vol. 6, p. 694.

In *The Origin of Forms and Qualities,* Boyle stated that the substantial forms of the Schoolmen provided an unintelligible explanation of the phenomena, whereas his own corpuscular hypothesis provided an account that was easily understandable.[52] Hence the hypothesis of the "vulgarly received notion of nature" of the Schoolmen could be rejected on the grounds (among others) that it was unintelligible. In *Free Inquiry into the Vulgarly Received Notion of Nature,* Boyle described this notion of the Schoolmen, who believed that

> nature is a most wise being, that does nothing in vain; does not miss of her ends; does always that, which (of the things she can do) is best to be done; and this she does by the most direct or compendious ways, neither employing any things superfluous, nor being wanting in things necessary; she teaches and inclines every one of her works to preserve itself: and, as in the microcosm, (man) it is she, that is the curer of diseases; so in the macrocosm (the world) for the conservation of the universe, she abhors a vacuum, making particular bodies act contrary to their own inclinations and interests, to prevent it, for the public good.[53]

This notion, Boyle complained, is

> so dark and odd a thing, that it is hard to know what to make of it, it being scarce, if at all, intelligibly proposed by them, that lay most weight upon it. For it appears not clearly, whether they will have it to be a corporeal substance, or an immaterial one, or some such thing, as may seem to be betwixt both; such as many Peripatetics do represent substantial forms, and what they call real qualities, which divers school-men hold to be (at least by miracle) separable from all matter whatsoever.[54]

His point was that supposed entities such as substantial forms must be either corporeal or incorporeal. If they are corporeal, they are not sufficient, Boyle thought, to account for the apparent intelligence attributed to nature. "A law," he claimed, referring to the laws of nature that the substantial forms allegedly obeyed, "being but a *notional rule of acting according to the declared will of a superior,* it is plain, that nothing but an intellectual Being can be properly capable of receiving and acting by a law."[55] On the other hand, if they are immaterial, the Schoolmen

52. Boyle, *Origin of Forms and Qualities, Works,* vol. 3, p. 5.
53. Boyle, *Notion of Nature, Works,* vol. 5, p. 174. On this essay, see Michael Hunter and Edward B. Davis, "The Making of Robert Boyle's *Free Enquiry into the Vulgarly Receiv'd Notion of Nature (1686),*" *Early Science and Medicine* 1 (1996), pp. 204–271.
54. Boyle, *Notion of Nature, Works,* vol. 5, p. 190.
55. Boyle, *Notion of Nature, Works,* vol. 5, p. 170.

offered no explanation of how immaterial substantial forms could affect a purely corporeal substance and cause such familiar phenomena as the ascension of water in pumps.[56] And if it be "a semi-substantia," which, Boyle noted, was what the Schoolmen themselves called substantial forms and real qualities, he rejoined that he would acknowledge "no such chimerical and unintelligible beings" as were both corporeal *and* incorporeal.[57]

Boyle did not consider the accounts the Schoolmen gave of substantial forms to merit the name of explications, for "to explicate a phænomenon, it is not enough to ascribe it to one general efficient" without going on to show exactly how the cause could produce the effect.[58] I show later in this chapter that Boyle indeed considered some of the phenomena of nature to be inexplicable by means of any available hypothesis, precisely because the exact way in which certain effects were produced was not known. The point here is that the Schoolmen were explicating phenomena on the grounds that nature somehow "knew" to act in particular ways, and this explication could be rejected, first of all, on the grounds that it was unintelligible.

Simplicity

It could also be rejected on the grounds that positing *any* power or intelligence to nature or some intermediary between God and the creation was superfluous, for in addition to being intelligible, an "excellent" hypothesis must be the simplest of all the good hypotheses that can be formulated to explain the phenomena – or at least it should contain nothing superfluous. Here again it is important to note that theological considerations lay not far from the surface. In his critique of the supposition of any such powers, intelligences, or other intermediaries, Boyle noted that he wished to speak both as a Christian and as a natural philosopher. As a Christian, his primary objection to any such entities was that they detract from God's omnipotence. As a Christian, Boyle believed that it was God's will and power that direct the course of

56. Boyle himself faced the same problem of explaining the power of an immaterial soul to move a human body, a problem he could not solve but could only characterize as "inexplicable."
57. Boyle, *Notion of Nature, Works*, vol. 5, p. 190. His *Origin of Forms and Qualities* also contains many objections to substantial forms on the grounds that they must be either material or immaterial, yet the Schoolmen deny that they are either.
58. Boyle, *Notion of Nature, Works*, vol. 5, p. 245.

nature, not any intermediary agent inherent in nature.[59] Further, in considering the lack of positive proof offered in support of the notion, Boyle noted that if the notion were true, one might expect so significant a truth as "God's grand vicegerent [sic] in the universe of bodies" to be found in scripture (especially in Genesis, which deals with the creation), whereas scripture is silent on the subject.[60] (Of course, scripture is similarly silent on the creation of atoms, the difference in the two cases apparently being that those who posited such intermediaries believed in the truth of their hypothesis despite what Boyle believed to be a total lack of empirical evidence, whereas nowhere, to my knowledge, did Boyle claim that his corpuscular hypothesis was true; further, because he conceived the atoms to be corporeal entities, direct evidence of their existence might be forthcoming with the development of more powerful microscopes.[61]) Another objection that Boyle raised as a Christian against the substantial forms of the Schoolmen was that the human soul was considered by them to be the substantial form of the body, yet substantial forms were generally considered to be perishable, hence posing a problem for the immortality of the soul.[62]

Although these objections are highly relevant to the question of the influence of Boyle's theology on his natural philosophy, my concern here is to examine Boyle's evaluation of the hypothesis on the secular basis of its superfluity. Noting that "a great part of the work of true philosophers has been, to reduce the true principles of things to the smallest number they can, without making them insufficient," Boyle saw no reason "why we should take in a principle, of which we have no need."[63]

Similarly, he objected to Henry More's appeal to a "hylarchic" princi-

---

59. Boyle's argument is summarized in Osler, "The Intellectual Sources of Robert Boyle's Philosophy of Nature," pp. 178–198.

60. Boyle, *Notion of Nature, Works*, vol. 5, p. 189. In *Things above Reason*, he explicitly rejected the idea that substantial forms might be "privileged things," and, as such, exempt from the rules of reasoning that apply to ordinary things. He offered no justification for this claim, but presumably it was based on the grounds that there is no scriptural revelation concerning substantial forms (*Works*, vol. 4, p. 451).

61. Catherine Wilson, *The Invisible World: Early Modern Philosophy and the Invention of the Microscope* (Princeton: Princeton University Press, 1995), esp. pp. 51–60.

62. Boyle, *Origin of Forms and Qualities, Works*, vol. 3, p. 40. In *Notion of Nature* (*Works*, vol. 5, p. 166), however, Boyle excluded the human soul from his critique of the system of the Schoolmen, as I noted in chapter 5.

63. Boyle, *Notion of Nature, Works*, vol. 5, p. 189.

ple on the same grounds. In the third edition of his *Antidote against Atheism* (1662) and in his *Enchiridion metaphysicum* (1671), More, a Cambridge divine, had appealed to Boyle's air-pump experiments in support of his own (More's) argument concerning the causal efficacy of a spirit (or "hylarchic principle") in the world. In 1672, Boyle responded to More in *An Hydrostatical Discourse, occasioned by the Objections of the Learned Dr. Henry More*.[64] Boyle's rejection of the hylarchic principle was relatively brief; he explicitly limited himself to a defense of his view that mechanical principles could account for the experimental results, and explicitly refrained from a direct attack on the views of the "learned Dr. *More*." Boyle's main point was that in his hypothesis of the spring and weight of the air, he was not attempting to "write a whole system, or so much as the elements of natural philosophy," but rather,

> having sufficiently proved, that the air we live in, is not devoid of weight, and is endowed with an elastical power or springiness, I endeavoured by those two principles to explain the phænomena exhibited in our engine.[65]

More might wish to attribute the spring and weight of the air to a "hylarchic principle," but Boyle considered himself "not obliged to treat of the cause of gravity in general,"[66] and hence not obliged to delve into the causes of the weight and springiness of the air. Nevertheless, Boyle insisted, in a lengthy passage that is worth quoting in full, that a good explanation should not contain anything superfluous:

> All that I have endeavoured to do in the explication of what happens among inanimate bodes, is to shew, that . . . the phænomena, I strive to explicate, may be solved mechanically, that is, by the mechanical affections of matter, without recourse to nature's abhorrence of a vacuum, to substantial forms, or to other incorporeal creatures. And therefore, if I have shewn, that the phænomena, I have endeavoured to account for, are explicable by the motion, bigness, gravity, shape, and other mechanical affectations of the small parts of liquors, I

---

64. For further discussion of this controversy, see Shapin and Schaffer, *Leviathan and the Air Pump*, pp. 207–224; R.A. Greene, "Henry More and Robert Boyle on the Spirit of Nature," *Journal of the History of Ideas* 23 (1962), pp. 451–474; John Henry, "Henry More versus Robert Boyle: The Spirit of Nature and the Nature of Providence," in *Henry More (1614–87): Tercentenary Studies*, edited by Sarah Hutton (Dordrecht: Kluwer Academic Publishers, 1990), pp. 55–76; and Alan Gabbey, "Henry More and the Limits of Mechanism," in *Henry More*, pp. 19–35.

65. Boyle, *Hydrostatical Discourse, Works*, vol. 3, p. 601.

66. Boyle, *Hydrostatical Discourse, Works*, vol. 3, p. 601.

have done what I pretended; which was not to prove, that no angel
or other immaterial creature could interpose in these cases; for
concerning such agents, all that I need say, is, that in the cases
proposed we have no need to recur to them. And this being agree-
able to the generally owned rule about hypotheses, that *entia non
sunt multiplicanda absque necessitate,* has been by almost all the
modern philosophers of different sects thought a sufficient reason to
reject the agency of intelligences, after *Aristotle,* and so many
learned men, both mathematicians and others, had for many ages
believed them the movers of the celestial orbs.[67]

Consistency

Another of Boyle's criteria for a "good" hypothesis was that it must be
consistent with related phenomena and not contradict any other known
phenomena, and he believed that the claim of the Schoolmen that nature
acts always to prevent a vacuum had been compromised by phenomena
produced in his experiments with the air pump. The Schoolmen argued
that when an inverted pipe was stopped up at one end, neither water nor
quicksilver descended, "lest it should leave a vacuum behind it." Yet
Boyle had shown that if an inverted tube "be but a finger's breadth
longer than 30 inches," quicksilver did indeed descend.[68] Similarly, the
"naturists" (as Boyle called the Schoolmen, because the term both described their notion of nature as an entity and distinguished them from
the new natural philosopher, the "naturalist")[69] claimed that nature has
implanted an appetite (that is, substantial forms) in heavy bodies that
causes them to descend to the center of the earth and an appetite in light
bodies to ascend toward heaven. This claim too was contradicted by
certain phenomena. A piece of wood, Boyle pointed out, if released in
air will sink toward the earth, in accordance with what the "naturists"
claimed, but if released in water will ascend with considerable force to
the top of the water, despite its alleged appetite for descent.[70] Of course,
the phenomena to which Boyle appealed could not *disprove* nature's
horror vacui or the appetite of heavy bodies to descend; horrors and
appetites could, perhaps, be thwarted. Boyle could not resist noting,

67. Boyle, *Hydrostatical Discourse, Works,* vol. 3, pp. 608–609. He also ap-
     pealed to *entia non sunt multiplicanda sine necessitate* in *Final Causes*
     (*Works,* vol. 5, p. 413).
68. Boyle, *Notion of Nature, Works,* vol. 5, pp. 193–194.
69. Boyle, *Notion of Nature, Works,* vol. 5, p. 168.
70. Boyle, *Notion of Nature, Works,* vol. 5, p. 194.

however, that the "naturists" would have quicksilver both descending in the tube because that was its natural tendency, as well as failing to descend in order to prevent a vacuum; one way or another, nature must "transgress her own ordinary laws."[71] In short, Boyle believed that his own (mechanical) hypothesis of the spring and weight of the air was consistent with experimental phenomena, whereas the "vulgarly received notion of nature" was much more difficult (if not impossible) to reconcile with experimental results.

Not only must a "good" hypothesis not contradict known phenomena, but an "excellent" hypothesis must be based on sufficient evidence, and neither the "vulgarly received notion of nature" of the Schoolmen nor the "hylarchic principle" (or "plastick nature") of the Cambridge Platonists was based on sufficient evidence. As Boyle put it in *Notion of Nature*,

> in matters of philosophy, where we ought not to take up any thing upon trust, or believe it without proof, it is enough to keep us from believing a thing, that we have no positive argument to induce us to assent to it, though we have no particular arguments against it.[72]

And the "naturists" offered no positive argument – at least not one based on incontrovertible evidence – for the efficacy of substantial forms. Appeals to such phenomena as a gardener's pot or inverted pipes certainly did not provide incontrovertible evidence in the face of other phenomena (including those resulting from Boyle's experiments). Boyle believed that if he had no reason to prefer the corpuscular hypothesis over the hypothesis of nature as a semi-deity, he would have no choice but to reject both, he said, and "endeavour to find some other preferable to both."[73] But his experimental program provided him with a storehouse of experimental evidence, all of which was consistent with the corpuscular hypothesis. The Schoolmen, on the other hand, maintained their belief in the efficacy of substantial forms in the face of evidence that was at best ambiguous, and at worst, recalcitrant.

The Question of The Falsity of Rejected Hypotheses

According to Boyle's criteria, then, a hypothesis should be rejected if it is unintelligible, if it violates Ockham's razor, if it contradicts known

---

71. Boyle, *Notion of Nature, Works*, vol. 5, p. 194.
72. Boyle, *Notion of Nature, Works*, vol. 5, pp. 188–189.
73. Boyle, *Notion of Nature, Works*, vol. 5, p. 246.

phenomena, or if it is based on insufficient evidence. To *reject* a hypothesis, however, is not the same thing as to declare it *false*. Certain of the criteria for a "good" and an "excellent" hypothesis would permit such a judgment; neither a "good" nor an "excellent" hypothesis, it will be recalled, could contain something impossible or manifestly false, nor could it suppose anything that is impossible or absurd, and it must be self-consistent. Nevertheless, declaring a hypothesis false was a judgment which Boyle made rarely, and then only where a particularly narrow hypothesis limited to extremely localized phenomena was manifestly false. This was the situation in which Boyle declared the "funicular hypothesis" of Franciscus Linus to be false.

In 1661, Linus published *Tractatus de corporum inseparabilitate* in which he attacked Boyle's *New Experiments Physico-mechanical, touching the Spring of the Air, and its Effects* (1660). Although in his work Linus agreed with Boyle that the air has a spring, he denied that the spring is sufficient to explain the apparently empty space in the tube in the Toricellian experiment. Linus, an Aristotelian, proposed that the space is actually filled with an invisible thread, a "funiculus," most likely consisting of rarefied mercury. In Boyle's *Defense of the Doctrine touching the Spring and Weight of the Air* (1662), he objected to Linus's hypothesis on the grounds that it was unintelligible and unnecessary, whereas Boyle's explanation of the empty space (the spring and weight of the air) was both intelligible and sufficient to explain the phenomena.[74] I mention this only in passing, however, because although being unintelligible and unnecessary provided grounds, in Boyle's opinion, for rejecting one hypothesis because a better one was available, the fact that a hypothesis was unintelligible and unnecessary was not grounds for declaring it false. In this case, however, Boyle *did* declare Linus's hypothesis false on the grounds that it was self-contradictory.

Boyle noted that Linus had argued that the phenomenon could not be explained any way other than by appeal to Aristotelian rarefaction, and that his explication of the funiculus as a rarefied substance that is "virtually extended" implied no contradiction. To this Boyle replied that

> if the affirming a body to be really and totally in this place, and at
> the same time to be really and wholly in another, that is, to be in
> this place, and not to be in this place, be not a contradiction, I know
> not what is.[75]

---

74. Boyle, *Defence of the Doctrine, Works*, vol. 1, pp. 162–163.
75. Boyle, *Defence of the Doctrine, Works*, vol. 1, p. 182. This explicit contradiction within a hypothesis should be distinguished from hypotheses that contain contradictory assertions, "it being very possible, that a man may

Further, Boyle rejected Linus's hypothesis on the grounds that he (Boyle) had proven his own hypothesis on the spring and weight of the air, and Linus's explication was inconsistent with Boyle's explication. Linus had admitted that there is a spring to the air, but had claimed that it was insufficient to "counterpoise" the column of mercury.[76] In rebuttal, Boyle told of two experiments made with a specially constructed J-shaped tube in which pressures that surpass atmospheric pressure could be produced. These experiments, which resulted in what was later to be called Boyle's Law, showed that "air being brought to a degree of density about twice as great as that it had before, obtains a spring twice as strong as formerly."[77]

Although Boyle was cautious in his claims for these experiments, noting that "in our table some particulars do not so exactly answer to what our formerly mentioned hypothesis might perchance invite the reader to expect," and that until he had made further trials he would "not venture to determine, whether or no the intimated theory will hold universally and precisely," nevertheless, "the trial already made sufficiently proves the main thing" – that the hypothesis of the spring and weight of the air was indeed sufficient to explain the phenomenon.[78] To posit, in the face of this evidence, a "funiculus" to make up for what Linus believed to be the insufficiency of the air's elasticity was not only superfluous, it contradicted Boyle's proof that the air's elasticity *was* a sufficient explanation.

It is significant, however, that the spring and weight of the air are relatively low on the "scale of causes."[79] As I have already mentioned, Boyle refused to ascend the "scale of causes" and ascribe a cause to the spring and weight of the air, but he was willing to speculate about possible mechanisms. Regarding the air's spring, he suggested that it could be accounted for in at least two ways. Each particle of air might itself be coiled, much like a fleece of wool. Or, the spring could result from the air particles being agitated by a fluid ether (that is, explained in terms of Cartesian vortices). Although Boyle considered the first possibility to be the one most easily understood, he saw no need, and indeed no

---

contradict himself in two several places of his works, and yet not be in both of them in the wrong" (*Origin of Forms and Qualities, Works*, vol. 3, p. 7).

76. Boyle, *Defence of the Doctrine, Works*, vol. 1, p. 124.
77. Boyle, *Defence of the Doctrine, Works*, vol. 1, pp. 156–57; quote, 157.
78. Boyle, *Defence of the Doctrine, Works*, vol. 1, p. 159.
79. Boyle discussed the advantages to be gained from limiting oneself at times to considerations of lower-level phenomena in *Experimental Essays, Works*, vol. 1, pp. 308–309.

way, to declare the account in terms of vortices (or any of the other possible explications) false. He considered it not to be his "business" "to assign the adequate cause of the spring of the air, but only to manifest, that the air hath a spring, and to relate some of its effects."[80] Linus, on the other hand, *did* ascribe a cause to his funiculus – the "fact" that nature abhors a vacuum.[81]

As unintelligible and superfluous as Boyle might find such hypotheses as nature's being an entity capable of "abhorring" and having "appetites," or the hypothesis of a "hylarchic principle," Boyle refused to pass a definitive judgment on such claims. This is most clearly seen in his controversy with Henry More when Boyle *explicitly acknowledged* that he could not show More's hypothesis of the "hylarchic principle" to be false. If, Boyle argued, he

> had been with those Jesuits, that are said to have presented the first watch to the king of *China,* who took it to be a living creature, I should have thought I had fairly accounted for it, if, by the shape, size, motion, &c. of the spring-wheels, balance, and other parts of the watch I had shown, that an engine of such a structure would necessarily mark the hours, though I *could not have brought an argument to convince the Chinese monarch, that it was not endowed with life.*[82]

On the "hylarchic principle," Boyle only observed that whether More's hypothesis "be true or no," such a principle "is at least not manifest."[83] Such a hypothesis might indeed be true, but in the absence

80. Boyle, *New Experiments, Works,* vol. 1, pp. 11–12, quote from p. 12. In his controversy with Hobbes, Boyle also explicitly refused to assign a cause for the spring and weight of the air; see *An Examen of Mr. Hobbes's "Dialogus Physicus de Naturâ Aëris," Works,* vol. 1, pp. 193–194. For Boyle's views on limiting explanations to the level of secondary qualities rather than accounting for phenomena in terms of primary qualities, see *Experimental Essays, Works,* vol. 1, pp. 308–310.

81. Boyle, *Defence of the Doctrine, Works,* vol. 1, p. 162. For a more detailed account of the controversy between Linus and Boyle see Shapin and Schaffer, *Leviathan and the Air Pump,* pp. 156–169. Shapin and Schaffer state that Boyle rejected Linus's funicular hypothesis on the grounds that it was "unintelligible" and unnecessary to account for the phenomena, which is true. However, they overlook the fact that Boyle also rejected the hypothesis on the grounds that it was *false* because it was itself inconsistent and also inconsistent with Boyle's hypothesis, which Boyle considered to have been experimentally established, and hence had become a matter of fact.

82. Boyle, *Hydrostatical Discourse, Works,* vol. 3, p. 627, emphasis added.

83. Boyle, *Hydrostatical Discourse, Works,* vol. 3, p. 617. For a similar interpretation of Boyle's position, see Sargent, *Diffident Naturalist,* p. 215.

of scriptural revelation on the creation of a "hylarchic principle," and in view of the fact that the data of experience would remain the same whether caused by the "hylarchic principle" or by matter and motion alone (a point I discuss more fully later), Boyle saw no reason to embrace such a hypothesis. As a natural philosopher, his rejection of the hypothesis was not related to its truth or falsehood, but rather to its violating his criteria for "good" and "excellent" hypotheses by being unintelligible and superfluous; as a Christian philosopher, his rejection of the hypothesis was due to his belief that it compromised God's omnipotence.

In fact, I have not found anywhere a sustained argument offered by Boyle that *any* of the three most general hypotheses considered by his contemporaries to be alternatives to the corpuscular hypothesis – the "hylarchic principle" or "plastick nature" of the Cambridge Platonists, the principles of Sulphur, Mercury and Salt of the Paracelsians, or the substantial forms of the Schoolmen – is *false*. Certainly Boyle *rejected* these hypotheses, but he did so on the grounds that none of them provided an intelligible explanation of *how* such an entity brought about the effects attributed to it, and hence did not offer any properly scientific explanation at all. The fact that a hypothesis was unintelligible and superfluous, however, did not make it false.

The Question of the Truth of the Corpuscular Hypothesis

The question naturally arises whether Boyle claimed the corpuscular hypothesis to be true. Just as I have found no sustained argument in which he declared that alternative hypotheses were false, neither have I found any in which he claimed that the corpuscular hypothesis was true. That he did not do so reveals a sophisticated understanding on Boyle's part as to exactly what such a claim would involve. To a certain extent, the differences between Boyle's corpuscular hypothesis and the various hypotheses championed by others were metaphysical in nature, and as such, incapable of confirmation or falsification. As I described here, when considering Henry More's claims for the agency of the "hylarchic principle," Boyle noted that whatever its ontological status might be, its activity, if any, was "not manifest" in the phenomena. Such entities as substantial forms and a "plastick nature" were not subject to detection by the senses. At the same time, the differences between the various hypotheses were not entirely metaphysical; corpuscles (or atoms), being specific material objects, might in time be detected by means of improved microscopes. Boyle, at least, believed that corpuscles were in principle detectable by the senses, although not all of his contemporaries agreed with him.[84] It should be noted, however, that even if the existence of corpuscles *did* come to be proven empirically, that would not entail the *nonexistence* of immaterial and imperceptible entities; indeed, it is difficult to conceive of any means by which they could be shown not to exist.[85] That Boyle was aware of this is shown by his comment about the Jesuits taking the first watch to China that was quoted in the previous section.

It is also the case that even if corpuscules were in fact detected by means of improved microscopes, their detection might present more questions than answers, for, as Boyle noted, solutions to the puzzles presented by nature often lead to "new difficulties, more capable than the first of baffling human understandings."[86] The fact is that Boyle believed that in creating the world, God was in no way bound to make that creation intelligible to human beings. Instead, God had created the world in whatever way he in his infinite understanding thought fit, leaving finite human understandings to speculate about that world "as well as they could."[87] In his own speculations about the created world, Boyle concluded that the corpuscular hypothesis was to be preferred because of its intelligibility, its simplicity, and its consistency with experimental results (not to mention its theological merits).

Advantages of the Corpuscular Hypothesis

In his rejection of the Scholastic hypothesis of nature as a semi-deity, Boyle stated, as I have noted, that if he had no reason to prefer the corpuscular hypothesis, he would have no choice but to reject them both and "endeavour to find some other preferable to both." But Boyle did prefer the corpuscular hypothesis for a number of reasons, some of which have become apparent in the course of this discussion. There were, first of all, his theological considerations, which should never

84. Boyle, *History of Colours, Works,* vol. 1, p. 680. For a succinct description of the argument from microscopy, see Christoph Meinel, "Early Seventeenth-Century Atomism: Theory, Epistemology, and the Insufficiency of Experiment," *Isis* 79 (1988), pp. 81–84. For the differing views of Boyle and Locke, see Walker, *The Invisible World,* esp. pp. 59–60 and 236–248.
85. Descartes faced a similar problem in his attempt to abolish substantial forms. See Daniel Garber, *Descartes' Metaphysical Physics* (Chicago: University of Chicago Press, 1992), pp. 110–111.
86. Boyle, *Things above Reason, Works,* vol. 4, p. 412.
87. Boyle, *Appendix to Christian Virtuoso, Works,* vol. 6, p. 694.

be underestimated. Indeed, such considerations were most likely the determining factors in his evaluation of alternative hypotheses, and I do not wish in any way to minimize the significance of Boyle's theological concerns. As I have stated repeatedly, it is my belief that his theological beliefs conditioned his entire worldview. My emphasis in this chapter, however, is Boyle's evaluation of hypotheses when he spoke purely as a natural philosopher rather than (as he so often did) as a *Christian virtuoso*. Speaking purely as a natural philosopher, he preferred the corpuscular hypothesis because it was more intelligible and hence easier to understand than were the alternative hypotheses, and it did not violate Ockham's razor by postulating superfluous entities. It was sufficient to explain a wide variety of phenomena (although not all, as I shall show later), and it did so without contradicting observed phenomena (as did nature's alleged abhorrence of a vacuum). In addition, it was preferable because it encouraged a deeper search into nature's secrets than did hypotheses that accounted for natural phenomena on the assumption of some power or intelligence on the part of "nature" that could not be investigated further. Others might claim that substantial forms or some other such entities caused observable phenomena, and end their explanations there, satisfied with an explanation that in fact explained nothing. But Boyle saw the role of the naturalist as investigating more deeply and inquiring into exactly what conditions brought about certain phenomena and what conditions did not.

Boyle's most concise published summary of these advantages may be found in *Of the Excellency and Grounds of the Mechanical Hypothesis* (1674). In this work, he added yet another advantage. He believed that a natural philosopher could both accept the corpuscular hypothesis *and* embrace any of a number of other hypotheses that were consistent with it. Such hypotheses, he claimed,

> if prudently considered by a skilful and moderate person, who is rather disposed to unite sects than multiply them, will be found, as far as they have truth in them, to be either legitimately (though perhaps not immediately) deducible from the mechanical principles, or fairly reconcileable to them.[88]

---

88. Boyle, *Excellency and Grounds of the Mechanical Philosophy, Works,* vol. 4, p. 72. For Boyle's irenicism in natural philosophy, see Barbara Beigun Kaplan, *"Divulging of Useful Truths in Physick": The Medical Agenda of Robert Boyle* (Baltimore: The Johns Hopkins University Press, 1993), esp. pp. 29, 56. See also John Henry, "Occult Qualities and the Experimental Philosophy: Active Principles in Pre-Newtonian Matter Theory," *History of Science* 24 (1986), p. 371.

The hypotheses that Boyle mentioned as being consistent with the corpuscular hypothesis included the four elements of the Aristotelians and the *tria prima* of the chymists.[89] Also included were the Platonic soul of the world, the "plastic power" (of the Cambridge Platonists), and the substantial forms of the Aristotelians.[90] Boyle's point was that these hypotheses involve either a material agent or an immaterial one. If material, then it was deducible from the corpuscular hypothesis, earth, for example, or mercury (with a lower-case "m") being reducible to matter and motion. If, on the other hand, the agent was immaterial, it must bring about its effects by means of making alterations in matter and motion. Even if "an angel himself should work a real change in the nature of a body," Boyle argued,

> it is scarce conceivable to us men, how he could do it without the assistance of local motion; since, if nothing were displaced, or otherwise moved than before, . . . it is hardly conceivable, how it should be in itself other, than just what it was before.[91]

In short, "by whatever principles natural things be constituted, it is by the mechanical principles, that their phænomena must be clearly explicated."[92]

It is worth noting here that Boyle also considered both his own mechanical hypothesis and that of Descartes to be consistent. Indeed, he had actually chosen the very term "corpuscular" because the corpuscular hypothesis as Boyle recommended it did not require making a judgment on the points of difference between Epicurean atomism and the Cartesian plenist mechanical philosophy. Both the Cartesians and the atomists "explicate[d] the same phænomena by little bodies variously figured and moved," and the hypothesis that these little bodies account for the phenomena did not require the natural philosopher to commit

89. Boyle, *Excellency and Grounds of the Mechanical Hypothesis, Works*, p. 72. For my use of the term "chymist," see pp. 129–130. For the consistency of Boyle's corpuscularianism with chemical matter theory, see Kaplan, "*Divulging of Useful Truths in Physick*," p. 59; Antonio Clericuzio, "A Redefinition of Boyle's Chemistry and Corpuscular Philosophy," *Annals of Science* 47 (1990), pp. 561–589.

90. Boyle, *Excellency and Grounds of the Mechanical Hypothesis, Works*, vol. 4, p. 72. Here he noted that "many of the judicious [philosophers]" considered such agents as the substantial forms to involve "self-repugnancy," but stopped short of expressing this as his own opinion.

91. Boyle, *Excellency and Grounds of the Mechanical Hypothesis, Works*, vol. 4, p. 73.

92. Boyle, *Excellency and Grounds of the Mechanical Hypothesis, Works*, vol. 4, p. 76.

himself to any position concerning the "notion of body in general, and consequently about the possibility of a true vacuum; [or] about the origin of motion, [and] the indefinite divisibleness of matter."[93]

Emphasizing the consistency of the corpuscular philosophy with its competitors was one of the ways in which Boyle facilitated its acceptance. If a fellow natural philosopher insisted on retaining one of these hypotheses (perhaps because that philosopher had his own theological motivations for clinging to, for example, the "plastick nature"), he could do so *and* embrace the corpuscular hypothesis.[94] And in time, Boyle believed, such philosophers would come to prefer the corpuscular hypothesis themselves, for the recent success of that hypothesis in the fields of "difficult phænomena" such as hydrostatics and optics spoke for itself. "When this philosophy is deeplier looked into and farther improved," Boyle claimed,

> it will be found applicable to the solution of more and more of the phænomena of nature. . . .
>
> In physical hypotheses, there are some, that, without noise, or falling foul upon others, peaceably obtain discerning men's approbation only by their fitness to solve the phænomena for which they were devised, without crossing any known observation or law of nature. And therefore, if the mechanical philosophy go on to explicate things corporeal at the rate it has of late years proceeded at, it is scarce to be doubted, but that, in time, unprejudiced persons will think it suffficiently recommended by its consistency with itself, and its applicableness to so many phænomena of nature.[95]

Boyle preferred the corpuscular hypothesis, then, for a number of reasons, including the rapid progress that had been made since it had been introduced and the fact that it was consistent with alternative hypotheses. In addition, its principles were, he thought, easy to conceive and its explications intelligible. Further, it fostered a deeper understanding of nature's secrets because it did not mislead the natural philosopher into thinking that a satisfactory cause of the phenomena had been assigned when the explication rested only on some unintelligible ad hoc appeal to the ability of nature to bring about the phenomena. It did not

93. Boyle, *Chymical Experiments, Works,* vol. 1, p. 355. He repeated this point in *Origin of Forms and Qualities* (*Works,* vol. 3, p. 7), where he explicitly noted that he was writing for "Corpuscularians in general, [rather] than any party of them."
94. I discussed Ralph Cudworth's theological motivation for his advocacy of the "plastick nature" in chapter 5.
95. Boyle, *Excellency and Grounds of the Mechanical Hypothesis, Works,* vol. 4, p. 77.

condone, as Allan Gabbey has so aptly expressed it, "the Spirit of the Causal Gaps."[96] Boyle's advocacy of the corpuscular hypothesis did not mean, however, that he believed that all natural phenomena could be satisfactorily explicated in terms of matter and motion.

Some Things Not Explicable By Any Means

Despite the advantages of the corpuscular hypothesis, Boyle was aware that it was incapable of providing an explanation for all phenomena. Some of his contemporaries, in arguing for the truth of some hypothesis or other, urged "that either the phænomenon must be explicated after the manner by them specified, or else it cannot at all be explicated intelligibly," on the assumption that any reasonable thinker would reject the possibility that a particular phenomenon was inexplicable. This was not an argument that Boyle accepted. First, he rejected the claim because "it is bold to affirm, and hard to prove, that what they cannot yet explicate by their principles, cannot possibly be explicated by any other men, or any other philosophy." More significantly, however, Boyle thought that even if the claim that no other intelligible explanation is possible were true, that would not provide a sufficient reason to accept the hypothesis in question:

> Supposing the argument to be conclusive, that either the proposed explication must be allowed, or men can give none at all, that is intelligible; I see not what absurdity it were to admit of the consequence. For who has demonstrated to us, that men must be able to explicate all nature's phænomena, especially since divers of them are so abstruse, that even the learnedest atomists scruple not to acknowledge their being unable to give an account of them.[97]

And indeed, because Boyle believed that God created the world commensurate to his own infinite understanding rather than human understanding, which God chose to limit (as I show in the next chapter), Boyle believed that some phenomena could not be intelligibly explicated. He made this clear in a passage in *A Free Inquiry into the Vulgarly received Notion of Nature* that parallels the passages just quoted. In this passage, Boyle acknowledged that mechanical principles could not satisfactorily explain all phenomena. And he went on to argue that

96. Gabbey, "Henry More and the Limits of Mechanism," p. 24.
97. Boyle, *Usefulness of Experimental Natural Philosophy, Works,* vol. 2, p. 46.

though mechanical principles could not be satisfactorily employed for explaining the phænomena of our world, we must not therefore necessarily recur to, and acquiesce in that principle, that men call nature, since neither will that intelligibly explain them; but in that case, we should ingenuously confess, that we are yet at a loss, how they are performed; and that this ignorance proceeds . . . from the natural imperfection of our understandings.[98]

That the natural philosopher could not explain all of the phenomena of nature, even by means of the corpuscular and mechanical hypothesis that Boyle believed to be superior to any other hypothesis was, in his opinion, all to the good because the inability to explicate all phenomena helped to prevent the natural philosopher's falling prey to the arguments of atheism. In *Usefulness of Experimental Natural Philosophy*, Boyle noted that he was aware that natural philosophy "is said to tempt to atheism, . . . by enabling men to give an account of the phænomena of nature, by the knowledge of second causes." To this, he responded that

it will not be so easy a matter, as many presume, for the contempla-tion of nature, to turn a considering man atheist. For we are yet, for aught I can find, far enough from being able to explicate all the phenomena of nature by any principles whatsoever.[99]

In this passage, Boyle specifically mentioned the generation of animals and the various phenomena exhibited by mercury to be inexplicable. Elsewhere, as I have shown at the beginning of this chapter, Boyle mentioned how the minute particles of matter cohere, how an immate-rial substance moves a material substance, and how bodies act on one another as being inexplicable phenomena.

Some phenomena, then, resist explication. And even when the natural philosopher was able to offer a good explanation of the phenomena, he was very often unable to show that it is a true explanation because it is often impossible for human beings to know the "true and genuine" causes of things. Although a natural philosopher might be able to show that the phenomena *might* be produced in a particular way, Boyle be-lieved that the naturalist often could not show that it was the *only* way the phenomena could be produced. He complained that

as confidently as many atomists, and other naturalists, presume to know the true and genuine causes of the things they attempt to explicate; yet very often the utmost they can attain to, in their explications, is, that the explicated phænomena may be produced

---

98. Boyle, *Notion of Nature, Works*, vol. 5, p. 246.
99. Boyle, *Usefulness of Experimental Natural Philosophy, Works*, vol. 2, p. 35.

> after such a manner, as they deliver, but not that they really are so. For as an artificer can set all the wheels of a clock a going, as well with springs as with weights; and may with violence discharge a bullet out of the barrel of a gun, not only by means of gunpowder, but of compressed air, and even of a spring: so the same effects may be produced by divers causes different from one another; and it will oftentimes be very difficult, if not impossible, for our dim reasons to discern surely, which of these several ways, whereby it is possible for nature to produce the same phænomena, she has really made use of to explicate them.

How could it be proved, Boyle asked, "that the omniscient God, or that admirable contriver, Nature, can exhibit phænomena by no ways but such as are explicable by the dim reason of man?"[100]

In support of this claim, Boyle appealed to Epicurus and to Aristotle, both of whom at least some places in their writings had limited themselves to proposing a plausible explication without insisting that any other explication would necessarily be false. Noting, however, that not all Epicureans and not all Aristotelians have followed this admirable modesty, Boyle added that "it is a very easy mistake for men to conclude, that because an effect may be produced by such determinate causes, it must be so, or actually is so."[101]

The Question of Progress in Natural Philosophy

Despite the excellence of the corpuscular hypothesis, it could not explain all of the phenomena of nature, at least not in Boyle's day. But the question of whether this inability was, in Boyle's opinion, inherent in the nature of things or because the corpuscular philosophy was in its embryonic stage remains to be considered. In some of the passages quoted in the preceding sections, Boyle used phrases that indicate that he thought human understanding had not *yet* been able to penetrate nature's secrets, and phrases to this effect appear in his writings far more often than the passages I have quoted indicate.

In the preface to *Origin of Forms and Qualities,* for example, he noted that the task of understanding how the motion, size, figure, and contrivance of the parts of matter (the corpuscles) might cause qualities

100. Boyle, *Usefulness of Experimental Natural Philosophy, Works,* vol. 2, pp. 45, 46.
101. Boyle, *Usefulness of Experimental Natural Philosophy, Works,* vol. 2, p. 45.

was in a very early stage. His purpose in writing the work was not to offer a full explanation of qualities, but simply to show that it was possible for the phenomena to be accounted for by a mechanical explanation alone, and that recourse to substantial forms was not necessary. By showing that the phenomena could be accounted for mechanically and by beginning a "history of qualities" (a record of experiments), Boyle hoped to encourage readers to

> contribute also their experiments and observations to so useful a work, and thereby lay a foundation whereon you, and perhaps I, may superstruct a more distinct and explicit theory of qualities than I shall at present adventure at.[102]

And in a passage in *High Veneration* in which he took note of the progress that had been made in the past, Boyle observed that he could

> scarce doubt, but by the farther improvement of telescopes posterity will have its curiosity gratified by the discovery both of new constellations, and of new stars in those, that are known to us already.[103]

Of perhaps even greater importance than progress made in understanding the origin of qualities and the contents of the heavens, however, was the progress to be made by means of the corpuscular hypothesis in all of those professions that "serve to provide men with food and rayment," such "husbandry, brewing, baking, fishing, fowling, and building."[104] In fact, Boyle devoted an entire section to experimental philosophy's "Usefulness to the Empire of Man over inferior Creatures" in his *Considerations touching the Usefulness of Experimental Natural Philosophy*. Where other practical concerns of his were involved, such as medicine, Boyle also anticipated future progress.[105]

Clearly, Boyle believed that progress made by future generations would increase man's store of knowledge. When asked explicitly about the extent of this future progress by one of the interlocutors in *Things above Reason*, his response was guarded. In that work, Eugenius asked,

102. Boyle, *Origin of Forms and Qualities, Works*, vol. 3, p. 14.
103. Boyle, *High Veneration, Works*, vol. 5, p. 152.
104. Boyle, *Usefulness of Experimental Philosophy, Works*, vol. 3, p. 402. For the relationship of natural philosophy to technology, see Larry R. Stewart, *The Rise of Public Science: Rhetoric, Technology, and Natural Philosophy in Newtonian Britain, 1660–1750* (Cambridge: Cambridge University Press, 1992); Michael Hunter, *Science and Society in Restoration England* (Cambridge: Cambridge University Press, 1981; Aldershot, Hampshire: Gregg Revivals, 1992), esp. pp. 87–112.
105. For Boyle's interest in medicine, see Kaplan, *"Divulging of Useful Truths in Physick,"* and, especially, Michael Hunter's "Boyle versus the Ga-

> may it not be well objected, that though the instances you have
> given have not been hitherto cleared by the light of reason; yet it is
> probable that they may be so hereafter, considering how great prog-
> ress is, from time to time, made in the discoveries of nature, in this
> learned age of ours?

To this, Boyle (Sophronius) responded that although

> this inquisitive age we live in will produce discoveries, that will
> explicate divers of the more hidden mysteries of nature, yet I expect,
> that these discoveries will chiefly concern those things, which either
> we are ignorant of for want of a competent history of nature, or we
> mistake by reason of erroneous prepossessions, or for want of free-
> dom and attention in our speculations. But I have not the like
> expectations as to all metaphysical difficulties, ... wherein neither
> matters of fact, nor the hypotheses of subordinate parts of learning,
> are wont much to avail.[106]

After noting that where human knowledge of God's nature is concerned,
Boyle thought that much would remain incomprehensible to human
understanding, he went on to discuss knowledge in general, saying that

> as mens inquisitiveness may hereafter extricate some of those grand
> difficulties, that have hitherto perplexed philosophers, so it may
> possibly lead them to discover new difficulties, more capable than
> the first of baffling human understandings. ... All which may ren-
> der it probable, that mens growing curiosity is not more likely to
> find the solutions of some difficulties, than to take notice of other
> things, that may prove more insuperable than they.

At this, Eugenius conceded that the question of how much progress
might be achieved was unanswerable because the question depended on
"future contingents."[107]

The appropriate question, then, is not whether Boyle thought human
understanding capable of progress in the uncovering of nature's secrets;
obviously he thought that such progress could and would be made. The
appropriate question is not "Will knowledge be increased?" but rather
"Are there limits beyond which human understanding will not be able
to pass?" In the next chapter, I show that Boyle did indeed believe that
there were such limits, limits that God, in his infinite wisdom, had seen
fit to impose on human understanding.

---

lenists: A Suppressed Critique of Seventeenth-century Medical Practice
and its Significance." *Medical History* 41 (1997).

106. Boyle, *Things above Reason, Works,* vol. 4, p. 412.
107. Boyle, *Things above Reason, Works,* vol. 4, pp. 412–413.

8

Boyle's Voluntarism and the
Limits of Reason

In *A Discourse of Things above Reason,* Boyle refused to commit himself
to any particular explanation as to exactly why he believed many of the
mysteries of Christianity and the secrets of nature to be impervious to
human understanding. As I showed in chapter 4, he suggested that the
difficulty might be due to our "lapsed state," or it might be related to
the way in which the human soul was united to the human body. But he
emphasized neither of these possibilities, being content most of the
time to appeal more generally to the incommensurability between the
understanding of finite beings and that of an infinite being. A careful
study of Boyle's writings as a whole, however, reveals that he believed
that God, as creator, had freely and deliberately chosen to limit the
rational faculties of human beings (and the rational faculties of angels,
as well). In creating the world commensurate with his own infinite
understanding and in limiting the rational faculties of created beings,
God had deliberately left "human understandings to speculate as well as
they could upon those corporeal, as well as other things."[1]

Boyle's belief in the incommensurability between finite created understandings and a world created in accordance with God's infinite understanding had significant implications for his conception of the task of
the natural philosopher. In chapter 7, I emphasized Boyle's refusal to
make knowledge claims about the truth or falsehood of the various
hypotheses available to explain the hidden mechanisms behind phenomena. There I stressed that although Boyle had reason to *prefer* the
corpuscular hypothesis on the grounds (inter alia) of its intelligibility, its
simplicity, its explanatory power, and its consistency with known matters of fact, he did not think himself justified in declaring such hypotheses as those involving the substantial forms of the Schoolmen or the
"plastick nature" of the Cambridge Platonists to be false. In the course
of that discussion, it became apparent that Boyle quite often appealed to

---

1. Boyle, *Appendix* to *Christian Virtuoso, Works,* vol. 6, p. 694.

the "dimness of our understandings" to explain why nature's ultimate secrets were remaining impenetrable.

In this chapter, I examine in detail Boyle's view that when God created the human mind, he deliberately chose to limit human understanding. Human beings are capable, in Boyle's opinion, of understanding exactly as much as God intended human beings to understand, and no more.

The Seventeenth-Century Background

In recent years, scholars investigating seventeenth-century natural philosophy have become increasingly aware of a significant relationship between a philosopher's conception of God and his conception of the proper way to go about attaining knowledge of the world God had created. The thesis, stated in its most simple form, is that those philosophers who emphasized God's power and will while paying less attention to his wisdom, goodness, and other attributes (that is, "theological voluntarists") tended to be empiricists, whereas those philosophers who emphasized God's wisdom, goodness, and other attributes besides his power and will (that is, "theological rationalists") tended to hold a rationalist theory of nature and of how to attain knowledge of nature. Theological rationalists, in other words, assumed that God had created the world according to reason, that the created world embodies at least some necessary relations, and that human reason is capable of discerning the nature of that creation either because of God-implanted innate ideas about the creation or because God had created the human intellect in such a way that it might discern the necessary relationships inherent in the creation. Theological voluntarists, on the other hand, believed that the created world is utterly contingent on God's will and that human beings can attain knowledge of what kind of world God had created only by investigating the world empirically.[2] Often, the thesis is ex-

2.   This thesis was originally formulated by M.B. Foster, "The Christian Doctrine of Creation and the Rise of Modern Natural Science," *Mind* 43 (1934), pp. 446–68, and "Christian Theology and Modern Science of Nature," published in two parts in *Mind* 44 (1935), pp. 439–66 and *Mind* 45 (1936), pp. 1–27. These essays have been reprinted, along with a number of other essays on the Foster thesis, in *Creation, Nature, and Political Order in the Philosophy of Michael Foster (1903–1959): The Classic "Mind" Articles and Others, with Modern Critical Essays*, edited by Cameron Wybrow (Lewiston: Edwin Mellon, 1992).

pressed in terms of a medieval distinction between two ways of understanding God's power in relationship to the created world – that is, between God's absolute power (*potentia Dei absoluta*) and his ordained power (*potentia Dei ordinata*), although the distinction was expressed at different times in different ways.

In one formulation, God's absolute power referred to what it was theoretically possible for God to have created (or refrained from creating) prior to the actual creation, whereas his ordained power referred to his governance of the created world in accordance with the particular laws of nature he had instituted. In another view, God's absolute power was not limited to the possibilities initially open to him at the creation, but referred as well to his occasional contravention in the usual course of nature (that is, to his power to perform miracles, just as a monarch has the power to contravene laws he himself has decreed). In either formulation, the emphasis was on God's power rather than his other attributes.[3]

When it comes to evaluating the views of individual philosophers, however, the thesis is more complex than might at first appear, and some confusion in the secondary "voluntarism" literature (as it has come to be called) has resulted from these complexities. Instead of the two rather clearly cut categories I have described, there are at least three different ways of categorizing the various positions, each of which can be expressed in terms of God's power and freedom (or lack thereof) in relation to various stages of the creation process.

Two of these positions are relatively straightforward. On the one

3. Francis Oakley, *Omnipotence, Covenant, and Order: An Excursion in the History of Ideas from Abelard to Leibniz* (Ithaca, NY: Cornell University Press, 1984), pp. 41–65. I have spoken of God as if he exists in the temporal order of things in order to clarify the concepts involved; I am grateful to Margaret J. Osler for having emphasized that strictly speaking, God is conceived as independent of the linear passage of time. For accounts of the medieval sources of voluntarist theology, see, in addition to Oakley, William J. Courtenay, "The Dialectic of Omnipotence in the High and Late Middle Ages," in *Divine Omniscience and Omnipotence in Medieval Philosophy*, edited by Tamar Rudavsky (Dordrecht: D. Reidel Publishing Company, 1985), pp. 243–269; Margaret J. Osler, *Divine Will and the Mechanical Philosophy: Gassendi and Descartes on Contingency and Necessity in the Created World* (Cambridge University Press, 1994), pp. 15–35. God's ordained power also refers to his faithfulness to promises made – to the covenantal nature of his relationship with the people of Israel. I am grateful to Edward B. Davis for this point.

hand, it was sometimes claimed in the seventeenth century that in creating this world, God was not free at all, but was instead bound by some standard, a standard that might be either external to himself or dictated by his own nature. On the other hand, it was sometimes claimed that God was, at the time of creation, entirely free to create any world whatsoever; moreover, it was *also* claimed that even *after* he had created this particular world, God remained utterly free to intervene in the world that he had created at any time that he might choose to do so.[4] The third position is less clear, and has posed problems of classification for scholars investigating voluntarism in the seventeenth century. In this third view, God was entirely free at the moment of creation, but once he had freely chosen to create this particular world, he was bound by his nature (his goodness, perhaps, or his immutability) to refrain from intervening in the world he had created.

George Rust

Certain of the Cambridge Platonists provide examples of the first position. It was the view expressed, for example, by George Rust (d. 1670), Bishop of Dromore in Ireland, in *A Discourse of the Use of Reason in Matters of Religion*.[5] In this work, Rust denied that *any* item of Chris-

---

4. The question in its simplest form dates back (at least) to Plato's *Euthyphro* (10a), in which Socrates asks, "Is what is holy holy because the gods approve it, or do they approve it because it is holy?" (*The Collected Dialogues of Plato,* edited by Edith Hamilton and Huntington Cairns [Princeton: Princeton University Press, 1961], p. 178).

5. George Rust, *A Discourse of the Use of Reason in Matters of Religion: Shewing, That Christianity Contains nothing Repugnant to Right Reason; Against Enthusiasts and Deists,* annotated by Henry Hallywell (London, 1683). This work was published posthumously. Hallywell, another Cambridge Platonist, translated it from Latin into English and included both texts in the published version. I consider it likely that Hallywell's publication of Rust's work was prompted at least in part by the publication of Boyle's *Things above Reason* in 1681, although nowhere in Hallywell's dedicatory epistle to fellow Cambridge Platonist Henry More or in his preface to the reader did he mention Boyle. Boyle knew himself to be subject to charges of enthusiasm as a result of the views he expressed in *Things above Reason;* see pp. 108–110. For a contemporary account of the Cambridge Platonist repudiation of voluntarism, see Edward Fowler, *The Principles and Practices, Of certain Moderate Divines of the Church of England, (greatly misunderstood) Truly Represented and Defended: wherein . . . some controversies, of no mean importance, are succinctly discussed . . .* (London, 1670),

tian belief is above reason (unless nothing more was meant by the claim than that the human intellect unaided by revelation is incapable of discovering some of the truths of Christianity revealed in scripture).[6] Each doctrine that has been expressly revealed and that Christians are required to believe must be something that can be understood, for "he that can persuade himself that he believes a thing that he does not understand, believes he knows not what."[7] Rust also made it clear that no revealed matter is contrary to reason as exercised by human beings – in direct contrast to Boyle's views as expressed in *Things above Reason*. Rust argued that if any revelation were to contradict reason, then either God is deceived, or God may deceive us, or there are no immutable and eternal relations of things, or our faculties are so framed as to deceive us even in those things that we perceive clearly and distinctly.[8]

It was in the process of affirming that there are immutable and eternal relations of things (clearly a rationalist position) that Rust noted there was one who had asserted "the Reasons of things to be contingent and Arbitrarious." To this, Rust exclaimed,

> but good God! What rash and abominable Positions do we hear! . . . You shall find no Heresie more deserving an *Anathema* then this. Nay, the very Jaws of Hell could not belch out any thing more detestable and blasphemous. For this robs God of his Wisdom, Immutability, Goodness, and all those other Perfections we attribute to him: It overthrows the Principles and Foundations of all Discourse; [and] makes contradictions become probable.[9]

---

esp. pp. 12–15. For a twentieth-century discussion, see Daniel A. Beck, "Miracle and the Mechanical Philosophy: The Theology of Robert Boyle in its Historical Context" (Ph.D. dissertation, University of Notre Dame, 1986), pp. 217–227.

6. Rust, *Use of Reason in Religion*, p. 26. Rust limited this claim to the items of belief expressly revealed in scripture.

7. Rust, *Use of Reason in Religion*, p. 26.

8. Rust, *Use of Reason in Religion*, p. 27.

9. Rust, *Use of Reason*, p. 27. In his annotations on Rust's work, Hallywell noted that the "certain person" (mentioned on p. 5 of the Latin version) was "Szydlovius, author of *Vindicia Quaestionum aliquot difficilium & controversarum in Theologia* printed at Franeker" (p. 51). That work, published in 1643, was by Jan Szydlowski, a professor at the Protestant school in Kiejdany (in present-day Lithuania). Of course, Boyle, too, asserted "the Reasons of things to be contingent and Arbitrarious." It is unlikely that Boyle was Rust's implicit target, however; Rust died in 1670, and most of Boyle's sustained discussions of voluntarism were published after that date. It does, however, seem likely that Hallywell's publication of

Although presumably Rust would have objected violently if someone had suggested that God is not omnipotent, it is clear that his emphasis was on God's "Wisdom, Immutability, [and] Goodness." It is also clear that Rust would not condone the elevation of God's omnipotence over these other attributes. Although God does not seem to have been bound to any standard external to himself, he was, according to Rust, bound by his own nature to create the world in such a way that it embodied those necessary relations inherent in his own eternal and immutable nature.

Joseph Glanvill

Joseph Glanvill, the Cambridge Platonist whose views on reason and religion were discussed in chapter 3, expressed this position more explicitly, if less colorfully, in *Lux Orientalis* (originally published in 1662; republished in 1682).[10] He too refused to elevate God's freedom and power over his wisdom and goodness, denying that God's "*meer will* is reason enough for his doing or forbearing any thing."[11] God's "*wisdom and goodness*," Glanvill continued, "are as it were the *Rules* whereby his *will* is directed."[12]

In another work, however, Glanvill expressed God's being bound at the time of creation in terms exemplifying the position that God was bound not by his own nature, but by a standard external to himself.

Rust's work in 1683 was occasioned by Boyle's publication in 1681 of *A Discourse of Things above Reason* and its accompanying *Advices in judging of Things said to transcend Reason*. Certainly Boyle had argued in *Things above Reason* and *Advices* that contradictions are indeed probable (from the perspective of finite human understanding).

10. Although educated at Oxford, Glanvill is generally regarded as having been influenced by the Cambridge Platonists. See, for example, Isabel Rivers, *Reason, Grace, and Sentiment: A Study of the Language of Religion and Ethics in England, 1660–1780*, vol. I, *Whichcote to Wesley* (Cambridge University Press, 1991), p. 28; see also Jackson I. Cope, *Joseph Glanvill: Anglican Apologist* (St. Louis: Washington University Press, 1956), pp. 85–86 and 126–143.

11. Glanvill, *Lux Orientalis*. In *Two Choice and Useful Treatises: The One Lux Orientalis; Or, An Enquiry into the Opinion of the Eastern Sages Concerning the Præexistence of Souls, being a Key to unlock the Grand Mysteries of Providence. The Other, A Discourse of Truth*, annotated by Henry More with a preface by Joseph Glanvill (London, 1682), p. 55. This work had originally been published some twenty years earlier, but had become scarce; hence its republication (unpaginated publisher's note).

12. Glanvill, *Lux Orientalis*, p. 56.

This passage occurs in his unpaginated preface to another of George Rust's works, *A Discourse of Truth* – a work that has been characterized as a "thoroughly intellectualist moral tract"[13] – which appeared in the same volume as Glanvill's *Lux Orientalis* in 1682, with both works annotated by Henry More.[14] In the preface, Glanvill explained that Rust's *Discourse of Truth* had been included in the volume because it had been circulating in manuscript and the copies were in danger of being corrupted, although here again it seems likely that the publication of Boyle's *Things above Reason* in 1681 prompted its inclusion, at least in part.[15] Expressing the hope that Rust's *Discourse* would contribute to a correct understanding of God's decrees in regard to the fate of human souls, Glanvill argued that a fundamental error is made when it is argued that

> things are good and just, because God wills them so to be. . . . If there be no settled Good and Evil, Immutable and Independent of *any* Will or Understanding, then God may have made his reasonable Creatures on purpose to damn them for ever.[16]

There is some uncertainty as to why Glanvill expressed his theological rationalism in these two different ways, the first of which depicts God as bound at the time of creation by his own nature, and the second as bound by a standard external to himself. Perhaps he expressed the latter conception in his preface to Rust's *Discourse of Truth* because it is the position he believed Rust to have held (and indeed it is possible to

---

13. John Henry, "Henry More versus Robert Boyle: The Spirit of Nature and the Nature of Providence," in *Henry More (1614–1687): Tercentenary Studies,* edited by Sarah Hutton (Dordrecht: Kluwer Academic Publishers, 1990), p. 63.

14. For More's own theological intellectualism, see John Henry, "Henry More versus Robert Boyle."

15. Glanvill's letter to the reader (unpaginated) introducing Rust's *A Discourse of Truth.* The full title of the volume is *Two Choice and Useful Treatises: The One Lux Orientalis; or An Enquiry into the Opinion of the Eastern Sages Concerning the Præexistence of Souls. Being a Key to unlock the Grand Mysteries of Providence. In Relation to Mans Sin and Misery. The Other, A Discourse of Truth, By the late Reverend Dr. Rust Lord Bishop of Dromore in Ireland. With Annotations on them both* [by Henry More] (London, 1682). One of Henry More's targets in his annotations to Rust's work was Richard Baxter, a theological voluntarist; for a discussion of More's conflict with Baxter, see John Henry, "A Cambridge Platonist's Materialism: Henry More and the Concept of Soul," *Journal of the Warburg and Courtauld Institutes* 49 (1986), pp. 182–189.

16. Glanvill's letter to the reader (unpaginated) introducing Rust's *A Discourse of Truth,* emphasis added.

interpret Rust in this way). What is important, however, is the fact that according to both versions of this position, God was not free to create just *any* kind of world. He was bound either by a standard external to himself, or by his own wisdom, goodness, or immutability to create this particular world, a world embodying "immutable and eternal relations of things," as Rust put it.[17]

Robert Boyle

Robert Boyle provides a paradigmatic example of the second position just described: That God, at the time of creation, was entirely free to create any world whatsoever, and further, that even *after* the creation, God remained utterly free to intervene in the world he had created at any time he might choose to do so. As Boyle succinctly expressed this position, "the laws of nature, as they were at first arbitrarily instituted by God, so, in reference to him, they are but arbitrary still."[18] In another passage, he elaborated on this, arguing that

> if we consider God as the author of the universe, and the free establisher of the laws of motion, whose general concourse is necessary to the conservation and efficacy of every particular physical agent, we cannot but acknowledge, that, by with-holding his concourse, or by changing these laws of motion, which depend perfectly upon his will, he may invalidate most, if not all the axioms and theorems of natural philosophy.[19]

Further, Boyle explicitly expressed the relationship between his theological voluntarism and his conception of the task of the natural philosopher when he explained that

> the world itself was first made before the contemplator of it, man: whence we may learn, that the author of nature consulted not, in the production of things, with human capacities; but first made things in such a manner as he was pleased to think fit, and afterwards left human understandings to speculate as well as they could upon those corporeal, as well as other things.[20]

---

17. A detailed analysis of Glanvill's empirical epistemology combined with his theological rationalism might reveal him to be a counterexample to the voluntarism thesis.
18. Boyle, *Appendix to Christian Virtuoso, Works*, vol. 6, p. 714.
19. Boyle, *Reason and Religion, Works*, vol. 4, p. 161.
20. Boyle, *Appendix to Christian Virtuoso, Works*, vol. 6, p. 694. In the same passage Boyle went on to point out that even if the world was not made by God but is eternal (as the ancient peripatetics taught) or was made by chance (as the Epicureans taught), "there is yet less reason to believe,

Boyle also held the typical voluntarist position that even after the natural philosopher has studied the world God actually chose to create and has reached certain conclusions about the nature of this world, those conclusions must remain provisional, for he considered them to be

> but limited, and such as I called collected truths, being gathered from the settled phaenomena of nature, and are liable to this limitation or exception, that they are true, where the irresistible power of God, or some other supernatural agent, is not interposed to alter the course of nature.[21]

I shall examine Boyle's voluntarism in more detail below. For the moment, I simply wish to emphasize that the two positions I have described – the obvious theological rationalism of Glanvill and Rust, and the obvious theological voluntarism of Boyle – are relatively unproblematic conceptually. It is the third position that has caused some confusion in the secondary literature.

René Descartes

In this third view, God was entirely free at the moment of creation, but once he had freely chosen to create this particular world he was bound by his nature (his immutability, perhaps, or his perfection) to refrain from intervening in the world he had created. René Descartes (1596–1650) provides an example of a seventeenth-century philosopher who held this position. According to Descartes, God could have created any world whatsoever. Prior to the creation, God was utterly free to pick and choose among all of the possible worlds. Probably the clearest expression of this aspect of Descartes' philosophy is his claim that the world could have been created in such a way that two plus three would *not* equal five.[22] Because of this freedom, some scholars have categorized Descartes as being a theological voluntarist.[23]

---

that there is any necessity that the nature of primitive things must be commensurate to our understandings; or that in the origin of other things any regard was had, whether they would or would not prove comprehensible to men."

21. Boyle, *Advices, Works*, vol. 4, p. 463.

22. René Descartes, *Meditations on First Philosophy*, in *The Philosophical Works of Descartes*, 2 vol., translated by Elizabeth S. Haldane and G.R.T. Ross (Cambridge: Cambridge University Press, 1967), vol. 1, p. 147.

23. R. Hooykaas, *Religion and the Rise of Modern Science* (Edinburgh: Scottish Academic Press, 1972), first American edition (Grand Rapids, MI: William B. Eerdmans Publishing Co., 1972), pp. 41–44.

This radical freedom, however, is only one aspect of Descartes's view of God's relationship to the created world: *After* he had chosen freely to create this particular world, God was bound by his immutability and perfection to refrain from intervening in the creation, if for no other reason than that altering the laws of nature would interfere with the certainty of the a priori knowledge of nature that God had guaranteed by means of "clear and distinct" (or innate) ideas. In short, once created, the laws of nature became necessary. The perfection of God's knowledge required that he not alter those laws because to do so would imply some alteration in his knowledge. Further, such a change would make God a deceiver because the innate ideas implanted in human minds would no longer be true. In short, God had freely bound himself to maintain the natural order he had freely chosen to create in order to guarantee the certainty of human knowledge, thereby introducing at least some necessity into the natural order. Because Descartes did not conceive of God as remaining free to intervene in the created world *after* the creation, he has also been characterized as a theological rationalist (or intellectualist).[24]

Clearly, the philosophy of Descartes cannot be characterized straightforwardly as either rationalist or voluntarist insofar as it contains elements of both.[25] As such, there is some confusion in the literature as to whether Descartes's philosophy provides a counterexample to the "voluntarist" thesis as defined at the beginning of this section (that theological rationalism is associated with a rationalist theory of knowledge, and that theological voluntarism is associated with an empiricist theory of knowledge). Margaret J. Osler, for example, classifying Descartes as a theological rationalist, believes that he fits the thesis quite well, whereas R. Hooykaas, who classifies Descartes as a theological voluntarist, believes that he provides a counterexample to the thesis.[26]

While acknowledging the complexity of Descartes's position, I think it more accurate to consider him a theological rationalist, primarily on the grounds that this is the way in which he was perceived by the most

24. Osler, *Divine Will and the Mechanical Philosophy,* esp. pp. 146–152.
25. Edward B. Davis, "God, Man and Nature: The Problem of Creation in Cartesian Thought," *Scottish Journal of Theology* 44 (1991), pp. 325–348. Davis argues that Descartes was a theological rationalist "who appealed to the divine will, paradoxically enough, primarily to further rationalist goals" (pp. 344–345), and that Descartes's appeal to the divine will was explicitly linked with contingency in nature. Davis's characterization of Descartes in this essay is more sophisticated than my brief description of it, and deserves careful study.
26. See notes 23 and 24.

thoroughgoing voluntarists of his time.[27] Pierre Gassendi (1592–1655), for example, considered Descartes to be a theological rationalist,[28] as did Blaise Pascal (1623–1662).[29] Robert Boyle also viewed Descartes as a rationalist. For example, he explicitly criticized Descartes' claim that the amount of motion in the world is necessarily constant because of God's immutability, arguing that the "properties and extent of the divine immutability are not so well known to us mortals."[30] Because God might well see fit to intervene in the created world, Boyle noted that he did not see why God's design might not be such "as might best be accomplished by sometimes adding to, and sometimes taking from, the quantity of motion He communicated to matter at first."[31] Despite differences of opinion as to exactly how Descartes should be characterized, however, the thesis that there was a historical connection between an emphasis on God's power and will and an empirical approach to attaining knowledge of nature, as well as a connection between an emphasis on God's wisdom, goodness, and other attributes (such as immutability) and a rationalist approach to knowing nature, has been received favorably.[32]

27. The complexity of Descartes' position is revealed by the fact that some of his contemporaries regarded him as a theological voluntarist – for example, Ralph Cudworth. See Francis Oakley, *Omnipotence, Covenant, and Order*, pp. 86–87.

28. Osler, *Divine Will and the Mechanical Philosophy*, esp. pp. 153–167. After Osler published *Divine Will and Mechanical Philosophy*, she concluded that "mitigated intellectualism" is the term that most appropriately describes the position held by Descartes (personal communication, September 1995).

29. Hooykaas, *Religion and the Rise of Modern Science*, pp. 44–46.

30. Boyle, *High Veneration, Works*, vol. 5, p. 140.

31. Boyle, *Final Causes, Works*, vol. 5, p. 396. Boyle developed this point more fully on the following page (397).

32. See, for example, in addition to works cited elsewhere in this chapter, Francis Oakley, "Christian Theology and the Newtonian Science: The Rise of the Concept of Laws of Nature," *Church History* 30 (1961), pp. 433–457); Eugene M. Klaaren, *Religious Origins of Modern Science* (Grand Rapids, MI: William B. Eerdmans Publishing Co., 1977); J. E. McGuire, "Force, Active Principles, and Newton's Invisible Realm," *Ambix* 15 (1968), pp. 154–208; James E. Force, "Newton's God of Dominion: The Unity of Newton's Theological, Scientific, and Political Thought," in *Essays on the Context, Nature, and Influence of Isaac Newton's Theology*, edited by James E. Force and Richard H. Popkin (Dordrecht: Kluwer Academic Publishers, 1990), pp. 75–102; Edward B. Davis, "Newton's Rejection of the 'Newtonian World View': The Role of Divine Will in

Specific Aspects of Boyle's Voluntarism

I turn now to a consideration of some particular aspects of Boyle's voluntarism, beginning with a consideration of the medieval roots of voluntarist themes and speculation as to particular influences on Boyle's voluntarist worldview.

Medieval and Reformation Emphasis on God's Power

There has been considerable scholarly debate concerning the history of theological voluntarism, and the tapestry of interwoven threads of thought is so complex that it will be impossible to do more than mention briefly some of the more significant aspects of that history here. Certainly the perceived conflict over the primacy of various of God's attributes had its origin in the clash between classical Greek rationality and the providential God of the Jews, a clash that reached its peak with the reception of the Aristotelian corpus into the Latin West. Despite the intellectual riches the Aristotelian system of natural philosophy had to offer, and attempts (such as those by Thomas Aquinas) to reconcile Aristotelianism with Christian theology, certain aspects of Aristotelianism (especially as interpreted by "The Commentator," the Spanish Muslim Ibn Rushd, or Averroës) were in fact inconsistent with Christian doctrine. The result was a series of official decrees beginning in 1210 and culminating in the condemnation of 219 propositions in 1277. Above all, the condemnations denied the natural necessity implicit in Averroist Aristotelianism and affirmed God's absolute power and freedom. As a result, many Christian philosophers began emphasizing the contingency of the created world and the inscrutability of God's will, a tendency that culminated in the nominalism of William of Ockham (c. 1285–1349) and his followers.[33]

Reformation theology provided a further emphasis on God's power,

---

Newton's Natural Philosophy," *Fides et Historia* 22 (1990), pp. 6–20; idem, "Christianity and Early Modern Science: The Foster Thesis Reconsidered," in *The Evangelical Engagement with Science*, edited by Mark Noll, David Livingstone, and Daryl Hart (New York: Oxford University Press, forthcoming); Betty Jo Teeter Dobbs, *The Janus Faces of Genius: The Role of Alchemy in Newton's Thought* (Cambridge: Cambridge University Press, 1991); Keith Hutchinson, "Supernaturalism and the Mechanical Philosophy," *History of Science* 21 (1983), pp. 297–333.

33. For the clash between Aristotelianism and Christianity and the outcome of the Condemnations, see the sources cited in chapter 1, fn 16.

although it should be remembered that Reformation theology was essentially Augustinian in nature, and that many Roman Catholics had held and continued to hold similar views regarding God's power even after the Reformation.[34] Both Martin Luther (1483–1546) and John Calvin (1509–1564) emphasized God's absolute sovereignty over the salvation of sinners. To think, as most medieval theologians had thought, that Christians could contribute to their own salvation by their actions attributed too much to human beings and too little to God; according to the Reformers, God's power and mercy were not contingent on human actions. Further, there is a certain symmetry between this view of God's active role in saving sinners and the relationship of created matter to its creator, with both matter and sinners being essentially passive and dependent on God as the source of activity, a view that may have facilitated the task of "baptizing" the mechanical philosophy of the seventeenth century. In any event, certainly the Reformation's emphasis on God's absolute power contributed at least to some extent to the voluntarism of such thinkers as Robert Boyle and Isaac Newton.[35]

Specific Influences

It is interesting to speculate about the influence of particular individuals on the development of Boyle's theological voluntarism, but it is difficult to trace direct influences with any degree of accuracy, primarily because little is known of his early life, and even less is known of the specific theological views held by many of the individuals who influenced his development. Still, much can be gleaned from the scattered sources that we do have. Almost certainly the most significant influence was that of his father, the great Earl of Cork, who was quick to credit God's direct intervention in his life for the events leading to his financial success, a tendency that eliminated his having to come to grips with his own ethically rather dubious means of amassing wealth.[36]

34. Davis, "Christianity and Early Modern Science."
35. See, for example, Gary Deason, "Reformation Theology and the Mechanistic Conception of Nature," in *God and Nature: Historical Essays on the Encounter between Christianity and Science,* edited by David C. Lindberg and Ronald L. Numbers (Berkeley and Los Angeles: University of California Press, 1986), pp. 167–191; and Edward B. Davis and Michael P. Winship, "Early Modern Protestantism," in *The History of Science and Religion in the Western Tradition: An Encyclopedia* (Garland Publishing Co., forthcoming).
36. Nicholas Canny, *The Upstart Earl: A Study of the Social and Mental World*

Another relatively early influence may have been Boyle's tutor, Isaac Marcombes (died c. 1654), who guided Boyle's studies from the time the youth left England in 1638 to tour the Continent (at the age of eleven) until he returned in 1644. Marcombes was a Frenchman by birth who made his home in Geneva, and was the nephew by marriage of Jean Diodati, the strict predestinationist. Boyle's extended tour included two stays in Geneva, each of which lasted almost two years (1639–1641 and 1642–1644).[37] During at least part of the first stay, two sections of Calvin's *Institutes* were read daily, and at some point while Boyle was in Geneva he met Diodati.[38] If indeed there is a link between Reformation theology and theological voluntarism, Marcombes may have been influential in the formation of Boyle's views, although too little is known of Marcombes's personal beliefs to be certain about this. It may be of some significance that originally Marcombes was recommended to Cork as a tutor for Boyle's older brothers by William Perkins (1558–1602), the Puritan author of *Armilla Aurea* (1590) and a staunch supralapsarian.[39]

Most likely, Boyle's voluntarist orientation was well-established by the time he returned to England from the Continent in 1644. If so, it would certainly have been reinforced by his contacts in the next few years with Samuel Hartlib (d. 1662) and his associates. Hartlib, John Dury (1596–1680), and Jan Comenius (1592–1670) were busy preparing for the soon-expected millennium with their plans to reform learning, unify

---

of *Richard Boyle, first Earl of Cork, 1566–1643* (Cambridge: Cambridge University Press, 1982), especially pp. 19–40; Terence O. Ranger, "Richard Boyle and the Making of an Irish Fortune," *Irish Historical Studies* 10 (1957), pp. 257–297; Malcolm Oster, "Biography, Culture, and Science: The Formative Years of Robert Boyle," *History of Science* 31 (1993), esp. pp. 180–183.

37. R.E.W. Maddison, "Studies in the Life of Robert Boyle, F.R.S. Part VII. The Grand Tour," *Notes and Records of the Royal Society* 20 (1965), pp. 51–77, and *The Life of the Honourable Robert Boyle, F.R.S.* (London: Taylor and Francis Ltd., 1969), pp. 45–56.

38. For the reading of Calvin, see Oster, "Biography, Culture, and Science," p. 190, citing a letter from Marcombes to Cork of February 1640. For meeting Diodati, see *Robert Boyle by Himself and His Friends*, edited by Michael Hunter (London: William Pickering, 1994), pp. xxvii, 26.

39. Oster, "Biography, Culture, and Science," p. 190; Maddison, *Life*, p. 21. For Perkins's theology, see Nicholas Tyacke, *Anti-Calvinists: The Rise of English Arminianism c. 1590–1640* (Oxford: Clarendon Press, 1987), esp. p. 29. A supralapsarian is one who believes that God's immutable decrees concerning the ultimate fate of individual souls predate Adam's fall.

Christians, and convert the Jews; their orientation was strongly providentialist.[40] Both Boyle and the members of the Hartlib circle were influenced by Johann Alsted (1588–1638), Comenius's professor of philosophy and theology at Herborn, whose massive *Encyclopædia* (Herborn, 1630) in turn reflected the influence of such Reformation thinkers as Luther and Melancthon.[41]

As interesting as it may be to speculate on the ways in which the young Boyle was exposed to voluntarist thought, any definitive explanation awaits more detailed studies on the theological orientation of those individuals who were in a position to share their views with the young Boyle. In the final analysis, however, what is remarkable is that his emphasis on God's freedom and power was simply a presupposition he never questioned. When Boyle encountered Pierre Gassendi's thought, he accepted Gassendi's voluntarism without hesitation.[42] Even though he also encountered theological rationalists such as Henry More and Joseph Glanvill, he apparently never questioned the correctness of his own presuppositions about God's nature and relationship to the created world. Boyle's God had providentially saved him from drowning at a very early age. While he was a student at Eton, his God had providentially saved him from being crushed when a wall in his room collapsed, as well as from being smothered by the dust caused by the collapse. In two other episodes while he was at Eton, God had, he believed, prevented his being killed in horse-related accidents.[43] It must have seemed perfectly obvious to Boyle that God not only could intervene in the

40. Charles Webster, *The Great Instauration: Science, Medicine and Reform, 1626–1660* (New York: Holmes and Meier, 1976), pp. 32–50; 67–77. See also Richard H. Popkin, "The Third Force in 17th Century Philosophy: Scepticism, Science and Biblical Prophecy," *Nouvelles de la république des lettres* 3 (1983), pp 35–63; idem, "Hartlib, Dury and the Jews," in *Samuel Hartlib and Universal Reformation: Studies in Intellectual Communication* (Cambridge: Cambridge University Press, 1994), pp. 118–136.

41. John Harwood, editor, *The Early Essays and Ethics of Robert Boyle* (Carbondale and Edwardsville: Southern Illinois University Press, 1991), Introduction, esp. pp. xxiv–xxxi.

42. Margaret J. Osler, "The Intellectual Sources of Robert Boyle's Philosophy of Nature: Gassendi's Voluntarism and Boyle's Physico-Theological Project," in *Philosophy, Religion, and Science, 1640–1700*, edited by Richard Kroll, Richard Ashcraft, and Perez Zagorin (Cambridge University Press, 1992), pp. 178–198.

43. Boyle, "An Account of Philaretus," in *Robert Boyle by Himself and His Friends*, pp. 4, 7.

world he had created, but in fact often did. And surely a God powerful enough to intervene in the creation was also a God powerful enough to have created any kind of world he chose.[44]

Appeals to Other Attributes

Regardless of the sources of Boyle's voluntarist orientation, there is no doubt that he provides a paradigmatic example of theological voluntarism, as has been shown by scholars such as Edward B. Davis, J.E. McGuire, Eugene M. Klaaren, Margaret J. Osler, and Timothy Shanahan.[45] However, Boyle's voluntarism did not preclude appeals to attributes of God's usually associated with theological rationalism (or intellectualism) such as God's wisdom and goodness. Almost all of the seventeenth-century philosophers were deeply rooted in the Christian tradition, and their philosophies were characterized by belief in a God who possessed *all* of the traditional attributes of wisdom, goodness, power, and so forth. Hence it is not surprising to find mention in their works of all of God's attributes, and Boyle is no exception.

Indeed, Boyle himself explained why certain of God's attributes are

44. According to Rose-Mary Sargent, Boyle's "voluntarist conception extended only to the creation of the world, not to its present state" (*The Diffident Naturalist: Robert Boyle and the Philosophy of Experiment* [Chicago: University of Chicago Press, 1995], p. 99). I believe this assessment to be mistaken, as evidenced not only by Boyle's explicit statement to the contrary in *Reason and Religion* (cited in note 19 above), but also by his belief that God had contravened the usual course of nature in his own life, his belief that spirits might intervene in alchemical processes (see chapter 6), and his investigation into the alleged healing powers of Valentine Greatrakes, the Irish "stroker" (see James R. Jacob, *Robert Boyle and the English Revolution: A Study in Social and Intellectual Change* [New York: Burt Franklin, 1977], pp. 164–176). See also J.J. MacIntosh, "Locke and Boyle on Miracles and God's Existence," in *Robert Boyle Reconsidered*, edited by Michael Hunter (Cambridge: Cambridge University Press, 1994), pp. 193–214.

45. Edward B. Davis, Jr., "Creation, Contingency, and Early Modern Science," pp. 122–181; Eugene M. Klaaren, *Religious Origins of Modern Science*, esp. pp. 85–184; J. E. McGuire, "Boyle's Conception of Nature," *Journal of the History of Ideas* 33 (1972), pp. 523–542; Margaret J. Osler, "The Intellectual Sources of Robert Boyle's Philosophy of Nature: Gassendi's Voluntarism and Boyle's Physico-Theological Project"; Timothy Shanahan, "God and Nature in the Thought of Robert Boyle," *Journal of the History of Philosophy* 26 (1988), pp. 547–569.

stressed in one context, while other attributes are stressed in other contexts. As he understood the situation, human understanding is not capable of framing

> a conception that shall comprehend the infinitely perfect nature of God in one single and simple idea; and therefore we are reduced to consider and represent him as it were in parts; contemplating him sometimes as omnipotent, and sometimes as wise, and sometimes as just, &c.[46]

In Boyle's case, the context in which he expressed his strongest and most striking statements of God's freedom and power was in passages in which he specifically considered God's relationship to the created world in general, many of which will be considered next. On the other hand, in those passages in which he emphasized God's wisdom and goodness, he was most often contemplating some specific aspect of that relationship, or recommending some particular attitude to his contemporaries. His most sustained discussion of God's wisdom, for example, can be found in his *Of the High Veneration Man's Intellect owes to God, peculiarly for his Wisdom and Power.* In this work Boyle's primary intention was to stress the inferiority of man's intellect as compared to God's, observing that

> it is not without some indignation, as well as wonder, that I see many men, and some of them divines too, who little considering what God is, and what themselves are, presume to talk of him and his attributes as freely and as unpremeditately . . . as if the nature and perfections of that unparalled [sic] Being were objects, that their intellects can grasp: and scruple not to dogmatize about those abstruse subjects.[47]

At least one of Boyle's targets in this work was Descartes, who had based his argument of the constancy of motion in the created world on God's immutability. The "properties and extent of the divine immutability," Boyle urged, "are not so well known to us mortals."[48] Another target was divines who had presumed to understand God's plan for the redemption of mankind, Boyle himself thinking that it could not be "an easy matter to have a mental eye so enlightened and so piercing, as to treat largely and worthily of so vast a subject" as that one, in which the "*wisdom*" of God was a "*mystery.*"[49] As Boyle put it,

> the distance betwixt the infinite creator and the creatures, which are but the limited and arbitrary productions of his power and will, is so

---

46. Boyle, *Appendix to Christian Virtuoso, Works,* vol. 6, p. 694.
47. Boyle, *High Veneration, Works,* vol. 5, p. 130.
48. Boyle, *High Veneration, Works,* vol. 5, p. 140.
49. Boyle, *High Veneration, Works,* vol. 5, p. 144.

vast, that all the divine attributes or perfections do by unmeasurable
intervals transcend those faint resemblances of them, that he has
been pleased to impress, either upon other creatures, or upon us
men."[50]

The emphasis on God's wisdom found in *High Veneration*, then, reflected Boyle's views on the *limits* of human reason, and not (as a
theological rationalist might claim) that the reason of human beings,
having been created in God's image, reflects the competence of divine
reason.

Another work in which repeated references to God's wisdom occur is
A Disquisition about the Final Causes of Natural Things. Here again,
Boyle had a particular point to make, specifically that the Cartesians and
Epicureans were wrong to exclude the study of final causes from the task
of the natural philosopher. In arguing the point, however, he was careful
to emphasize that although it was "erroneous to deny, that any thing
was made for ends investigable by man," he was in no way claiming
that human understanding is capable of comprehending fully God's
purposes in the creation; to do so would be "a presumption."[51]

In short, it is not surprising to find passages in which a voluntarist
emphasized God's wisdom, nor those in which a rationalist emphasized
God's omnipotence. What is relevant is the relative emphasis placed on
various of God's attributes in the philosopher's writings as a whole and
the contexts within which passages relating to God's attributes occur.
Perhaps the context in which Boyle's voluntarism is most striking is one
that has not been explored systematically – his conception of God's
intentions when he created the human intellect.

God's Will and Human Reason

Boyle's voluntarism affected not only how he thought knowledge of the
created world ought to be attained, but also the limits that he believed
God had set on human reason.[52] He suggested, for example, that

on divers occasions it may be a usefull supposition to Imagine such
a state of things as was that which proceeded the beginning of the

---

50. Boyle, *High Veneration, Works,* vol. 5, pp. 148–149.
51. Boyle, *Final Causes, Works,* vol. 5, pp. 396; 397.
52. In "Creation, Contingency, and Early Modern Science," Edward B. Davis
    noted (passim) that Boyle was a voluntarist where God's creation of human
    understanding is concerned, but he neither developed nor argued the point
    fully.

Creation for since then there was no being besides God himselfe who is Eternall all Beings of what kind soever that had a beginning must derive those natures & all their faculties from his Arbitrary will.[53]

Immediately after having made this point, he went on to add,

& consequently man himselfe & all intellectual as well as all Corporeall Creatures, were but just such as he thought fit to make them. And as he freely establisht the affections of schemetisms & the Laws of Motion by which the universe was framed, & doth act; so he freely constituted the Reason of Man & other created intellects.[54]

Boyle expressed this view more fully in another unpublished essay, as well as in the posthumously published *Appendix to the First Part of the Christian Virtuoso*. In the former, he noted that

thô among his other Creatures he [God] was pleasd to frame some that were of an Intelligent Nature, as Angells & Rational Souls; & to endow these with Intellectual faculties, more or less capacious & inlightned, as he thought fit; yet he reservd to himself, as his prerogative, the Intimate & Compleat Knowledge of his Works; allowing to Humane Minds (to confine our present Discourse to that sort of Intellectual Beings) various degrees of knowledge, according *to* their congenit Aptitudes for knowledge; & *to* the several Circumstances, that from time to time They find themselves in.[55]

In this passage, Boyle went on to say that just as perfect knowledge is, in God, accompanied by "a full Complacency & Satisfaction," so will

53. Boyle Papers, vol. 36, fol. 46v. Perhaps Boyle meant "preceded" rather than "proceeded."

54. Boyle Papers, vol. 36, fol. 46v.

55. Boyle Papers, vol. 8, fol. 187. Somewhat surprisingly, the passage I have quoted is immediately preceded by one of the very few passages in which Boyle emphasizes God's reason in a way that makes him sound more like a theological rationalist (of the type that considered God bound at the time of creation to create the world in accordance with his own eternal ideas) than a voluntarist, claiming that created things are "but Transcripts of those Eternal *Ideas* of the divine mind, according to which they were fram'd, and their Conformity to which induc'd their Author, upon the review of his six dayes works, to look upon all the things that he had made as (in their respective kinds) *very good*." I do not know whether Boyle considered this view to be in some way reconcilable with the extreme voluntarism he expressed elsewhere. The folio is headed, "A summary Account of the Sentiments, upon which the following Essay is mainly founded, may be thus represented," and is in the hands of two of Boyle's amanuenses (Robin Bacon and Thomas Smith). I am grateful to an anonymous reader for having corrected an erroneous handwriting attribution.

God reward humans who use their skill and industry to increase their knowledge insofar as they are able to do so with "correspondent measures of Delight, & Contentment."[56]

In the *Appendix*, first published in 1744, Boyle repeated his view that when God

> was pleased to create other beings, as he did it freely and uncompelled, so he gave to each of those he created such a nature, and such determinate faculties and powers as he pleased. And of the beings, which he created intelligent, among which are the rational minds of men, he endowed each sort or order with such a measure or degree of the intellectual faculty, as he thought fit, for the ends and purposes for which in his infinite wisdom he created them; but to none of these did he impart a knowledge boundless as his own, that infinite knowledge being the essential prerogative of God, and not communicable to any mere creature.[57]

Further, Boyle noted that the relationship of the human mind to the created world affects the ability of human beings to reason about that world. Because God created the world before he created man, he did not take human understanding into consideration when he created the world. Instead, he created the world as he thought fit, leaving human beings to speculate about that world as best they can. As Boyle put it,

> I fear we men have too good a conceit of ourselves, when we think that no such thing can have an existence, or at least have a nature or being, as we are not able to comprehend. For if we believe God to be the author of things, it is rational to conceive that he may have made them commensurate, rather to his own designs in them, than to the notions we men may best be able to frame of them. . . . The world itself was first made before the contemplator of it, man: whence we may learn, that the author of nature consulted not, in the production of things, with human capacities; but first made things in such a manner as he was pleased to think fit, and afterwards left human understandings to speculate as well as they could upon those corporeal, as well as other things.[58]

---

56. Boyle Papers, vol. 8, fol. 187.
57. Boyle, *Appendix* to *Christian Virtuoso, Works*, vol. 6, p. 697. As Boyle put it in *Advices*, "to discover particular truths is one thing, and to be able to discover the intercourse and harmony between all truths, is another thing" (*Works*, vol. 4, p. 466).
58. Boyle, *Appendix* to *Christian Virtuoso, Works*, vol. 6, p. 694. Here, Boyle went on to point out that even if the world had not been made by God but is eternal (as the ancient peripatetics taught) or was made by chance (as the Epicureans taught), "there is yet less reason to believe, that there is any necessity that the nature of primitive things must be commensurate to our

He made the same point in *Usefulness of Experimental Natural Philosophy*:

> If God be allowed to be, as indeed he is, the author of the universe, how will it appear that he, whose knowledge infinitely transcends ours, and who may be supposed to operate according to the dictates of his own immense wisdom, should, in his creating of things, have respect to the measure and ease of human understandings; and not rather, if of any, of angelical intellects?[59]

In short, Boyle believed that God imposed whatever limits to understanding he thought fit, and had most likely chosen to grant greater intellects to angels than to human beings. He also believed that God, in creating individual human beings, chose to endow some with "a greater measure of intellectual abilities" than others.[60] In *Style of the Scriptures,* for example, he noted that the scriptures were written both for individuals "with elevated and comprehensive intellects" and for individuals "very weak and illiterate."[61] God had given "to the intelligent productions of his power and will various degrees of intellectual capacities," but he had reserved full understanding for himself.[62]

Boyle's voluntarism should be understood to include not only God's freedom to produce this natural world rather than some other, and not only God's power to intervene in the created world, but also Boyle's belief that in creating the human intellect, God freely chose to limit its capacity for understanding.[63] In Boyle's view, God, had he so chosen, could have engineered things in such a way that the human intellect would have been capable of an a priori knowledge of the natural world – but he did not choose to do so.

---

understandings; or that in the origin of other things any regard was had, whether they would or would not prove comprehensible to men" [as noted also in fn 20].

59. Boyle, *Usefulness of Experimental Natural Philosophy, Works,* vol. 2, p. 46.
60. Boyle, *Excellency of Theology, Works,* vol. 4, p. 26.
61. Boyle, *Style of the Scriptures, Works,* vol. 2, p. 262.
62. Boyle, *Things above Reason, Works,* vol. 4, p. 466.
63. In arguing that Boyle's voluntarism had a minimal influence on his science, Daniel A. Beck misses this point completely, arguing that there were a number of *other* theological justifications for Boyle's empiricism, including the limits of human understanding ("Miracle and the Mechanical Philosophy: The Theology of Robert Boyle in its Historical Context," Ph.D. dissertation, Notre Dame, 1986, pp. 185–186). It is also a point missed by Richard S. Westfall in his otherwise excellent *Science and Religion in Seventeenth-Century England* (New Haven: Yale University Press, 1958), a work published when the significance of Boyle's voluntarism was only beginning to be recognized by scholars.

The Christian Virtuoso's Final Reward

Boyle thought that God set such limits to human understanding because he believed that God, in his infinite wisdom, has reserved a full understanding (both of revealed truths and of nature) for the afterlife. In *Seraphic Love*, for example, he noted that

> in heaven our faculties shall not only be gratified with suitable and acceptable objects, but shall be heightened and enlarged, and consequently our capacities of happiness as well increased as filled. ... Our then enlarged capacities will enable us, even in objects which were not altogether unknown to us before, to perceive things formerly undiscerned, and derive thence both new and greater satisfactions and delights.[64]

> There, probably, we shall satisfactorily understand those deep and obscure mysteries of religion, ... [and] we shall discern not only a reconcileableness, but a friendship, and perfect harmony, betwixt those texts, that here seem most at variance.[65]

After death, "all that unwelcome darkness, that here surrounded our purblind understandings, will vanish at the dawning of that bright, and ... eternal day; wherein the resolution of all those difficulties, which here exercised (and perhaps distressed) our faith, shall be granted us to reward it."[66] Indeed, Boyle found this ultimate reward a strong motivation to renounce temptations "for which inconsiderate mortals are wont to forfeit the interest their Saviour so dearly bought them."[67]

Not only will an understanding of theological mysteries be the Christian's reward in heaven, but the Christian natural philosopher's knowledge of nature will be increased as well; in fact, the increased understanding with which God will reward him will be proportionate to the amount of effort which he has put into his attempts to understand nature while on earth. Boyle made this clear in *The Second Part of the Christian Virtuoso*, saying that

> the study of nature, with design to promote piety by our attainments, is useful, not only for other purposes, but to increase our

---

64. Boyle, *Seraphic Love, Works*, vol. 1, p. 283.
65. Boyle, *Seraphic Love, Works*, vol. 1, p. 289.
66. Boyle, *Seraphic Love, Works*, vol. 1, p. 290.
67. Boyle, *Seraphic Love, Works*, vol. 1, p. 290. Similar claims concerning the perfection of the human intellect in the afterlife may be found in *Reason and Religion, Works*, vol. 4, p. 161, *Excellency of Theology, Works*, vol. 4, p. 32, and *Usefulness of Experimental Natural Philosophy, Works*, vol. 2, p. 33.

knowledge, even of natural things, if not immediately, and presently, yet in time, and in the issue of affairs. For, at least, in the great renovation of the world, and the future state of things, those corporeal creatures, that will then, be knowable, notwithstanding such a change, as the universe will have been subject to, shall probably be known best by those, that have here made their best use of their former knowledge, which there will, together with their other gifts, be *congruously,* as well as *munificently,* rewarded. And then the attainment of a high degree of knowledge, which here, was so difficult, may, to the enlightened and enlarged mind, become as *easy,* as it will be *satisfactory:* and our improved understandings, will, with joy, perceive, how much all the knowledge, that we can give ourselves of God's works, is inferior to what their divine Author can impart to us, of them.[68]

Because God created the world to be commensurate with his own infinite understanding, and because he deliberately circumscribed the understanding of the intelligent creatures which he created, human understanding is incompetent to understand fully nature's secrets. Only in the afterlife might the "divine Author" of both nature and man increase the inferior understanding of human beings. How wise it was, Boyle thought, for God to have created such a powerful incentive to be both a Christian and a Virtuoso.

68. Boyle, *Christian Virtuoso, Second Part, Works,* vol. 6, p. 776. In another passage in the same work (pp. 788–789), Boyle observed that it is possible that the "*new heaven, and new earth,*" might be so totally different from the present state of affairs that the knowledge of nature obtained in this present life might be useless. Here his point was to show superiority of theological knowledge to natural knowledge, for even if our knowledge of nature must undergo a radical transformation in order to understand the nature of the new state of affairs, or even if a knowledge of nature might be irrelevant in heaven, our knowledge of divine things will be "retained, [and] highly improved" in the afterlife. For Boyle's eschatology, see Malcolm Oster, "Millenarianism and the New Science: The Case of Robert Boyle," in *Samuel Hartlib and Universal Reformation,* pp. 137–148.

Conclusion

In the preceding chapters, I have shown how Boyle's views on reason's limits in the context of theology were reflected in his views on the limits of reason in the context of natural philosophy. The entire point of this book, indeed, has been the extent to which Boyle *expected* to find parallels between what might be known about the secrets of revelation and what might be known about the secrets of nature. Boyle approached his studies of theology and nature with certain assumptions about reason's limits, and then interpreted his findings in such a way as to reinforce those assumptions: His God had deliberately chosen to limit the power and scope of human reason, leaving human beings in something of a state of perpetual blindness concerning the ultimate truths of both nature and Christianity.

This picture of Boyle differs markedly from our traditional picture of him as a confident and single-minded advocate of both the corpuscularian philosophy and rational religion. The image of Boyle as emphasizing a more rational approach to both nature and theology than he in fact did has its roots in the way he was portrayed in the early and highly selective accounts of his life given by Gilbert Burnet in Boyle's funeral sermon on the second of January, 1692, and by Thomas Birch in the biography of Boyle accompanying the first edition of his *Works* (1744).[1] This view of him has been perpetuated since by scholars who have taken the canonical image of Boyle at face value, finding it all too easy to elide various aspects of his thought and personality that violate twentieth-century conceptions of religious and scientific rationality. In closing, I discuss how it is that previous scholars have failed to recognize the extreme limits Boyle placed on human reason. In addition, I discuss some of the broader implications of my own study of his epistemology, with particular attention to claims made in recent years by scholars who

---

1. Michael Hunter, Introduction to *Robert Boyle by Himself and His Friends*, edited by Michael Hunter (London: William Pickering, 1994).

have approached Boyle studies from the point of view of social and political history.

One reason for scholars' having misconstrued Boyle's views on reason's limits has been the disproportionate amount of attention paid to his writings about natural religion compared with his writings about the relationship of reason to the content of revelation. In *Robert Boyle on Natural Philosophy*, for example, Marie Boas Hall, whose writings on Boyle have been tremendously influential, discussed his theology without even mentioning *Things above Reason* or revelation.[2] Indeed, in discussing Boyle's literary style, she claimed that *The Sceptical Chymist* was the only work published in dialogue form – a rather remarkable oversight in light of the fact that both *Things above Reason* and its accompanying *Advices* were written in the form of dialogues.[3]

Another reason for misinterpretations of Boyle's views on reason's limits stems from scholars' having neglected to study his writings on things above reason in the context of the threat of Socinianism and of the controversies between Calvinists and Arminians concerning predestination. It had been assumed that Boyle's *Things above Reason* was a response to Spinoza's *Tractatus* (1670, reprinted 1677).[4] Of course, Spinoza may well have been *one* of Boyle's targets, but the evidence I have presented in chapters 2 and 3 shows that he also had other targets in mind. As I pointed out in the Introduction, failing to read Boyle's *Things above Reason* in the light of arguments presented by his contemporaries makes it all too easy to take his claim that nothing in scripture is contrary to reason at face value. After all, were Spinoza Boyle's only target, the argument that there are indeed things *above* reason would have done the job quite well. When *Things above Reason* is read in the light of the Socinian and predestination controversies, however, it becomes clear that Boyle's argument included the claim that although revelation may *contradict* finite human understanding, nothing in scripture is contrary to *God's infinite* understanding. Realizing this makes it possible to understand for the first time the extent to which Boyle in fact limited human understanding.

2. Marie Boas Hall, *Robert Boyle on Natural Philosophy: An Essay with Selections from His Writings* (Bloomington: Indiana University Press, 1965), esp. pp. 47–51.
3. Hall, *Robert Boyle on Natural Philosophy*, p. 39.
4. Lotte Mulligan, "Robert Boyle, 'Right Reason', and the Meaning of Metaphor," *Journal of the History of Ideas* 55 (1994), p. 241; Rosalie Colie, "Spinoza in England 1665–1730," *Proceedings of the American Philosophical Society* 107 (1963), pp. 195–196.

In short, the rationality of Boyle's thought has been greatly exaggerated. In 1918, for example, S.G. Hefelbower claimed that "Boyle held that God had given man reason by which he could know the principles of natural religion, but that this was not enough. By reason we know that there are things above reason, which are not contradictory to it."[5] In 1945, M.S. Fisher argued that in Boyle's view, "reason is to give way only to admit revelation, upon revelation's entry reason is to return."[6] In 1955, R.M. Hunt claimed that "Boyle avowed that no true Scriptural contradictions could be found."[7] Even Richard S. Westfall, whose detailed treatment of Boyle's theological views is both sophisticated and for the most part correct, made the same error. "Since absolute truths are the principles upon which assent to revelation itself must be grounded," Westfall stated, Boyle believed that "any revelation that seems to contradict them must be interpreted in a way that removes the difficulty."[8] This is the Socinian position – not Boyle's. Boyle did indeed claim that a revealed truth might be contrary to reason in that either it is self-contradictory or it contradicts some conclusion rightly reached by unaided reason. As I have shown in chapter 4, in such cases Boyle was content to acknowledge both truths on the assumption that God, in his infinite understanding, comprehends the harmony among all truths. It is

5.	S.G. Hefelbower, *The Relation of John Locke to English Deism* (Chicago: University of Chicago Press, 1918), p. 73.
6.	M.S. Fisher, *Robert Boyle, Devout Naturalist: A Study in Science and Religion in the Seventeenth Century* (Philadelphia: Oshiver Studio Press, 1945), p. 126; see also p. 155. In making this claim, Fisher relied on Boyle's *Appendix* to the *Christian Virtuoso, Works*, vol. 6, p. 712. There is reason to believe that Boyle intended this passage to reflect the views of an interlocutor opposed to Boyle's own views on reason's limits; see my discussion of the reliability of the *Appendix* in chapter 4.
7.	R. M. Hunt, *The Place of Religion in the Science of Robert Boyle* (Philadelphia: University of Pittsburgh Press, 1955), p. 73.
8.	Richard S. Westfall, *Science and Religion in Seventeenth-Century England* (New Haven: Yale University Press, 1958), p. 173. Westfall grounded his claim on the same dubious passages in the *Appendix*; see note 6. Similar assessments are made by John Redwood in *Reason, Ridicule and Religion: The Age of Enlightenment in England, 1660–1750* (Cambridge: Harvard University Press, 1976), p. 211; G.A.J. Rogers, "Boyle, Locke, and Reason," *Journal of the History of Ideas* 27 (1966), pp. 213–214; and Mulligan (see note 13 below). Steven Shapin's assessment of Boyle's position, that is, that there "were wholly legitimate limits to the coherence and consistency that might be demanded of our body of beliefs" (*A Social History of Truth: Civility and Science in Seventeenth-Century England* [Chicago: University of Chicago Press, 1994], p. 230), is correct, although Shapin does not seem to recognize the significance of this for Boyle's thought as a whole.

only when Boyle's views about things contrary to reason are understood correctly that the extent to which he circumscribed reason's power and scope can be recognized.

Reading Boyle's works on things above reason in the context of the theological controversies of his day also raises questions about the nature of his relationship to the Church of England. While there is no doubt that Boyle in fact conformed to the Church of England after its reestablishment, there is considerable evidence of his support, financial and otherwise, of those who refused to conform.[9] The relationship of his thought on reason's limits to that of certain of the nonconformists (which I discussed in the first part of this book) provides additional evidence that he was not as straightforward an advocate of the Anglican church as he has been depicted. Further, doubt is raised as to what extent he should be considered to reflect the thought of latitudinarians.[10]

Interconnections between "right reason," rational Christianity, natural religion, the new philosophy, and latitudinarianism have been taken for granted, and Boyle has been considered a paradigmatic example of this general line of thought. Barbara Shapiro, for example, who has so effectively demonstated correlations between natural philosophy and theology in the establishment of a probabilistic epistemology in seventeenth-century England, stated that latitudinarians "firmly believed in the compatibility of natural and revealed religion and the rationality of both."[11] Boyle figures prominently in her account, and her portrayal of him is for the most part accurate – except for the fact that Boyle's

9. See, in addition to chapters 3 and 4, Jan W. Wojcik, "The Theological Context of Boyle's *Things above Reason*," in *Robert Boyle Reconsidered*, edited by Michael Hunter (Cambridge: Cambridge University Press, 1994), pp. 139–155; Hunter, Introduction to *Robert Boyle by Himself and His Friends*, esp. pp. lxx–lxii.

10. When scholars from a number of disciplines convened at the Clark Library in Los Angeles in 1987 to explore the nature of English latitudinarianism, "four days of often intense discussion yielded surprisingly little substantive or methodological agreement about the putative object of [their] pursuit," and for this reason no reference to "latitudinarianism" was made in the title of the volume of essays resulting from that meeting (Richard Kroll, Introduction to *Philosophy, Science, and Religion in England 1640–1700* [Cambridge: Cambridge University Press, 1992], p. 2). See also my review of this book in *Journal of the History of Philosophy* 32 (1994), pp. 141–142.

11. Barbara J. Shapiro, *Probability and Certainty in Seventeenth-Century England: A Study of the Relationships between Natural Science, Religion, History, Law, and Literature* (Princeton, NJ: Princeton University Press, 1983), p. 95.

views on the relationship of reason to revelation are not adequately distinguished from those of such contemporaries as Joseph Glanvill and John Locke. Those differences may well have been irrelevant to Shapiro's subject matter. Nevertheless, the portrayal of Boyle as an exemplar of latitudinarianism is misleading, and helps to perpetuate the received historiography of Boyle's views on reason.

Lotte Mulligan has explicitly described Boyle's views on "right reason" as his "fifty-year-long defense of the Church of England" in an essay in which Boyle is depicted as a paradigmatic example of a post-Restoration mechanical philosopher who retained and indeed emphasized the need to subjugate the conclusions of natural reason to the higher truths of scriptural revelation in the face of the tendency of contemporaries to rely increasingly on natural reason. As far as the general thesis goes, she is quite correct. Unfortunately, she misinterpreted Boyle's views on reason's limits, and misleadingly concluded that his writings on reason and revelation constituted a defense of "the Church of England and its brand of 'rational religion'."[12]

It may seem trivial for me to insist that Boyle's writings on reason constituted a life-long defense of the Christian religion rather than a defense of the Church of England. It may seem trivial to insist on the *extreme* limits Boyle placed on human reason in view of the fact that almost all commentators acknowledge that he placed the "mysteries of the faith in a realm where they may remain unscathed because untouched" by human reason.[13] But both points are crucial. In recent years, a number of scholars have approached Boyle studies from the point of view of social and political history, and by having taken for granted the received historiography that links Boyle with latitudinarianism and rational religion they have made claims about his goals and

12. Mulligan, "Robert Boyle, 'Right Reason,' and the Meaning of Metaphor," quotations from pp. 240 and 257; on p. 243 Mulligan claimed that for Boyle, the basic beliefs of Christianity were the same as the doctrines of the Church of England. Mulligan has erred too in her assessment of Boyle and his category of things "contrary to reason" by failing to recognize the distinction that Boyle made between theology and natural philosophy where the "unsociable" category of things above reason is concerned, arguing that Boyle "was showing that there were many things – not only in theology but also in natural philosophy – which were 'above reason' but in no way 'contrary' to it" (ibid., p. 246); she too based her interpretation on dubious passages in the *Appendix* (see fn 8).

13. Eugene M. Klaaren, *Religious Origins of Modern Science: Belief in Creation in Seventeenth-Century Thought* (Grand Rapids, MI: Eerdmans, 1977), p. 132.

motives that are unsupported by the evidence, misleading, or only partially correct.

Linking Boyle with latitudinarianism is the error made by James R. Jacob and Margaret C. Jacob in a series of books and articles in which they claim that the latitudinarians represented a conservative force in seventeenth-century England, aligned with the establishment position and dedicated to the advancement of trade, industry, and empire, and that Boyle exemplified this position.[14] To whatever extent that this may have been true of the latitudinarians generally, the claim definitely oversimplifies the complexity of Boyle's views, particularly in the assumptions made by the Jacobses about Boyle's purposes and motivations. No doubt Boyle did have much in common with the latitudinarians. He shared their views on the importance of securing a natural theology as a basis for belief in revealed Christianity and their advocacy of "moral certainty" as an epistemically reliable category of knowledge.[15] And insofar as latitudinarianism may have been associated with toleration for dissenters, here too Boyle was in agreement, although his motives for doing so reflected his concern with liberty of conscience rather than the advantages of toleration to the state.[16] The Jacobses' error lies not in assuming that there were areas in which Boyle and the latitudinarians shared similar beliefs, but rather in the assumption that whatever might be claimed of the latitudinarians might be also claimed of Boyle.

A similar error is to be found in Steven Shapin and Simon Schaffer's *Leviathan and the Air Pump.* In this study of Boyle's actual experimental

14. See inter alia, James R. Jacob, *Robert Boyle and the English Revolution: A Study in Social and Intellectual Change* (New York: Burt Franklin, 1977), esp. p. 158; James R. Jacob and Margaret C. Jacob, "The Anglican Origins of Modern Science: The Metaphysical Foundations of the Whig Constitution," *Isis* 71 (1980), pp. 251–267.

15. The best sources for these aspects of Boyle's thought remain Westfall, *Science and Religion in Seventeenth-Century England*; Shapiro, *Probability and Certainty.*

16. The extent to which latitudinarians consistently advocated toleration between the Restoration and the Glorious Revolution is, in any event, questionable. See, for example, John Marshall, "The Ecclesiology of the Latitude-men 1660–1689: Stillingfleet, Tillotson and 'Hobbism'," *Journal of Ecclesiastical History* 36 (1985), pp. 407–427; Jan W. Wojcik, "'Behold the Fear of the Lord': The Erastianism of Stillingfleet, Wolseley, and Tillotson," in *Heterodoxy, Spinozism and Free Thought,* edited by S. Berti, F. Charles-Daubert, and R. H. Popkin (Dordrecht: Kluwer Academic Publishers, 1966), pp. 357–374.

practice, Shapin and Schaffer claim that the dispute between Boyle and Hobbes was a dispute over how knowledge is to be constructed, with Boyle advocating publicly verified "matters of fact" as epistemically privileged and as logically distinct from hypotheses. Further, Shapin and Schaffer claim that the "experimental philosophers aimed to show those who looked at their community an idealized reflection of the Restoration settlement."[17] *Leviathan and the Air Pump* is a valuable addition to studies in the history of philosophy of science in that the authors raise questions about the role of human agency in the construction of scientific knowledge that are of critical importance in our understanding of the relationship of science to the political goals and aims of its practitioners. Nevertheless, the study poses a number of problems.

First, there was not, in Boyle's view, as sharp a demarcation between matters of fact and hypotheses as the authors claim.[18] Second, the assumption that there was any consensus of political goals and aims among the experimental philosophers (a category that Shapin and Schaffer equate with the Royal Society) is unfounded.[19] Finally, the question of the extent to which Boyle himself supported the Restoration settlement is itself a questionable one, as I have already mentioned.

The Boyle portrayed by such scholars as the two Jacobses and Shapin and Schaffer is, in a sense, a dogmatic Boyle – a Boyle who had clearly defined goals of aligning experimental science and religious toleration with the restored crown and the reestablished church, which in turn were concerned with propagating trade and empire. Such a view of Boyle owes much to the long history of his being portrayed as an advocate of rational religion and rational science. Scholars approaching

17. Steven Shapin and Simon Schaffer, *Leviathan and the Air Pump: Hobbes, Boyle, and the Experimental Life* (Princeton, NJ: Princeton University Press, 1985), p. 341.
18. See my discussion in chapter 6; see also Rose-Mary Sargent, *The Diffident Naturalist: Robert Boyle and the Philosophy of Experiment* (Chicago: University of Chicago Press, 1995), esp. pp. 131–135.
19. Michael Hunter, "Latitudinarianism and the 'Ideology' of the Early Royal Society: Thomas Sprat's *History of the Royal Society* (1667) Reconsidered," in idem, *Establishing the New Science: The Experience of the Early Royal Society* (Woodbridge, Suffolk: Boydell Press, 1989), pp. 45–71; idem, *The Royal Society and its Fellows 1660–1700* (Chalfont St. Giles: British Society for the History of Science, 1982). As Sargent has pointed out, Shapin and Schaffer frequently use comments made by others as support for the position they claim Boyle held, most often when they link his pneumatic experiments with sociopolitical aims (*Diffident Naturalist,* pp. 9, 223 n. 48.

Boyle studies from the perspective of the sociology of science have taken at face value the portrait of Boyle drawn by previous generations of Boyle scholars. And the portrait of Boyle depicted by past generations of Boyle scholars has failed to do justice to the complexity of his thought, neglecting or misinterpreting as it has various aspects of Boyle's thought and personality that violate twentieth-century conceptions of religious and scientific rationality.

I am not denying that Boyle *perceived himself* as having a rational approach to both science and religion. Even as he argued reason's limits he offered rational grounds for doing so, and certainly he perceived himself as an exemplar of levelheaded and moderate right-thinking. Seeing Boyle this way, however, is only part of the picture, and is misleading, for such a picture fails to do justice to the tentativeness (indeed, the "diffidence")[20] of his approach to both the secrets of nature and those of theology. It fails to convey the extent to which Boyle's entire worldview was conditioned by his theological beliefs and the absolute primacy of his theological assumptions. Among these assumptions was a deep-seated conviction that God had deliberately chosen to limit human understanding, and this conviction conditioned his approach to the controversies in theology, natural philosophy, and politics that provided the background for seventeenth-century philosophy in England. This conviction led to an extreme modesty in his dealings with those who held alternative views in theology or natural philosophy, and a marked disinclination to pronounce himself to be a possessor of truth. Certainly Boyle's God rewarded those who sought knowledge, but exactly how much his God had intended human beings to understand was not clear to him. Advances in natural knowledge would be accompanied, Boyle thought, by new mysteries. Man's desire to know could be only partially satisfied in this life.

20. I have borrowed this word, of course, from the apt title of Rose-Mary Sargent's *The Diffident Naturalist.*

Bibliography

Boyle's Works Cited

The following is a list of the works I have cited that were published in *The Works of the Honourable Robert Boyle*, edited by Thomas Birch, 6 vols. (London, 1772, reprinted with an introduction by Douglas McKie, Hildesheim: Georg Olms, 1965). I have listed first the shortest title by which I have referred to an individual work in the text. This is followed in parentheses by the full title of the work and the date it was first published. References to manuscript sources are given in the notes and are not repeated here.

Advices (*Advices in judging of Things said to transcend Reason*, 1681).

An Examen of Mr. Hobbes's "Dialogus Physicus de Naturâ Aëris" (*An Examen of Mr. T. Hobbes's Dialogus Physicus de Naturâ Aëris, as far as it concerns Mr. Boyle's Book of New Experiments touching the Spring of the Air, &c., with an Appendix touching Mr. Hobbes's Doctrine of Fluidity and Firmness*, 1662).

Appendix to Christian Virtuoso (*Appendix to the first Part of the Christian Virtuoso*, 1744).

Chymical Experiments (*Some Specimens of an attempt to make Chymical Experiments useful to illustrate the notions of the Corpuscular Philosophy*, 1661).

Christian Virtuoso (*The Christian Virtuoso, shewing that by being addicted to Experimental Philosophy a Man is rather assisted than indisposed to be a good Christian: The First Part*, 1690).

Christian Virtuoso, Second Part (*The Christian Virtuoso, the Second Part*, 1744).

Considerations Touching Experimental Essays (*A Proæmical Essay, wherein, with some Considerations touching Experimental Essays in general, is interwoven such an Introduction to all those written by the Author, as is necessary to be used for the better understanding of them*, 1661).

Defence of the Doctrine (*A Defence of the Doctrine touching the Spring and Weight of the Air, proposed by Mr. Robert Boyle in his new Physico-mechanical Experiments, against the Objections of Franciscus Linus, wherein the Adversary's Objections against the Elaterists are examined*, 1662).

Essays of the Strange Subtilty (*Essays of the Strange Subtilty, Great Efficacy, Determinate Nature of Effluviums*, 1673).

Excellency and Grounds of the Mechanical Hypothesis (Some occasional Thoughts about the Excellency and Grounds of the mechanical Hypothesis, 1674).

Excellency of Theology (The Excellency of Theology compared with Natural Philosophy, (as both are Objects of Men's Study) discoursed of in a Letter to a Friend, 1674).

Final Causes (A Disquisition about the final Causes of natural Things, wherin is inquired, whether, and (if at all) with what Cautions a Naturalist should admit them, 1688).

High Veneration (Of the high Veneration Man's Intellect owes to God, peculiar for his Wisdom and Power, 1684).

History of Colours (Experiments and Considerations touching Colours, 1663).

History of Fluidity and Firmness (The History of Fluidity and Firmness, 1661).

Hydrostatical Discourse (An Hydrostatical Discourse occasioned by the Objections of the learned Dr. Henry More against some Explications of new Experiments made by Mr. Boyle, 1672).

New Experiments (New Experiments Physico-mechanical, touching the Spring of the Air, and its Effects; made, for the most part, in a new Pneumatical Engine, 1660).

Notion of Nature (A Free Inquiry into the Vulgarly Received Notion of Nature, 1686).

Origin of Forms and Qualities (Origin of Forms and Qualities according to the Corpuscular Philosophy, illustrated by Considerations and Experiments, written formerly by way of Notes upon an Essay upon Nitre, 1666).

Possibility of the Resurrection (Some Physico-theological Considerations about the Possibility of the Resurrection, 1675).

Reason and Religion (Some Considerations about the Reconcileableness of Reason and Religion, 1675).

Reflections upon a Theological Distinction (A Discourse about the Distinction that represents some Things as above Reason, but not contrary to Reason, 1690).

Sceptical Chymist (The Sceptical Chymist; or Chymico-Physical Doubts and Paradoxes touching the Experiments, whereby Vulgar Spagyrists are wont to endeavour to evince their Salt, Sulphur and Mercury to be the true Principles of Things, 1661).

Seraphic Love (Some Motives to the Love of God, pathetically discoursed of in a Letter to a Friend, 1659).

Specimens of Chymical Experiments (Some Specimens of an Attempt to make Chymical Experiments useful to illustrate the Notions of the Corpuscular Philosophy, 1661).

Style of the Scriptures (Some Considerations touching the Style of the Holy Scriptures, extracted from several Parts of a Discourse concerning divers Particulars belonging to the Bible, written divers Years since to a Friend, 1661).

Things above Reason (A Discourse of Things above Reason, inquiring whether a Philosopher should admit there are any such, 1681).

Usefulness of Experimental Natural Philosophy (Some Considerations touching the Usefulness of Experimental Natural Philosophy, proposed in a familiar Discourse to a Friend, by way of Invitation to the Study of it: Parts I and Part II, Section I, 1663).

Usefulness of Experimental Natural Philosophy, the Second Part, the Second Section (cited by full title, 1671).

Other Works Cited

Adams, Robert Merrihew. "Where Do Our Ideas Come From? – Descartes vs. Locke." In *Innate Ideas,* edited by Stephen P. Stitch. Berkeley and Los Angeles: University of California Press, 1975.

Alexander, Peter. *Ideas, Qualities and Corpuscles: Locke and Boyle on the External World.* Cambridge: Cambridge University Press, 1985.

Aquinas, St. Thomas. *Faith, Reason and Theology: Questions I-IV of his Commentary on the "De Trinitate" of Boethius,* translated with introduction and notes by Armand Maurer. Toronto: Pontifical Institute of Mediaeval Studies, 1987.

——*On There Being One Intellect,* translated by Ralph McInerney. In Ralph McInerney, *Aquinas Against the Averroists: On There Being Only One Intellect.* West Lafayette, IN: Purdue University Press, 1993.

——*Summa Theologica I.* In *Basic Writings of Saint Thomas Aquinas,* edited by Anton C. Pegis. 2 vols. Random House: New York, 1944, vol. 1.

Aristotle. *Rhetoric,* translated by W. Rhys Roberts. In *The Basic Works of Aristotle,* edited by Richard McKeon. New York: Random House, 1941.

Armstrong, Brian G. *Calvinism and the Amyraut Heresy: Protestant Scholasticism and Humanism in Seventeenth-Century France.* Madison: University of Wisconsin Press, 1969.

Barlow, Thomas. "The Case of a Toleration in Matters of Religion." In *Several Miscellaneous and Weighty Cases of Conscience.* London, 1692.

Baxter, Richard. *The Arrogancy of Reason against Divine Revelations, Repressed; or, Proud Ignorance the cause of Infidelity, and of Mens Quarrelling with the Word of God.* London, 1655.

——*The Judgment of Non-Conformists, of the Interest of Reason, in Matters of Religion.* London, 1676.

——*The Reasons of the Christian Religion.* London, 1667.

——*The Unreasonableness of Infidelity: Manifested in four discourses.* London, 1655.

Beck, Daniel A. "Miracle and the Mechanical Philosophy: The Theology of Robert Boyle in its Historical Context." Ph.D. dissertation, University of Notre Dame, 1986.

Biddle, John. *A Twofold Catechism: The One simply called A Scripture-Catechism; the Other, A brief Scripture-Catechism for Children; wherein the*

Chiefest points of the Christian Religion, being Question-wise proposed, resolve themselves by pertinent Answers taken word for word out of the Scripture, without either Consequences or Comments. Composed for their sakes that would fain be Meer Christians, and not of this or that Sect, inasmuch as all the Sects of Christians, by what names soever distinguished, have either more or less departed from the simplicity and truth of the Scripture. London, 1654.

Birch, Thomas. "The Life of the Honourable Robert Boyle." In *The Works of the Honourable Robert Boyle,* edited by Thomas Birch. 6 vols. London, 1772. Reprinted with an introduction by Douglas McKie. Hildesheim: Georg Olms, 1965.

Boas, Marie. See Hall, Marie Boas.

Boulton, Richard. *The Theological Works of the Honourable Robert Boyle, Esq., Epitomiz'd.* 3 vol. London, 1715.

Budgell, Eustace. *Memoirs of the Lives and Characters of the Illustrious Family of the Boyles; Particularly of the Late Eminently Learned Charles Earl of Orrery. . . . With a Particular Account of the famous Controversy between the Honourable Mr. Boyle, and the Reverend Dr. Bentley, concerning the Genuineness of PHALARIS'S Epistles; also the same translated from the Original Greek. With an Appendix Containing the Character of the Honourable Robert Boyle Esq; Founder of an Annual Lecture in Defence of Christianity. By Bishop Burnet, and others. Likewise his Last Will and Testament.* London, 1737.

Burnet, Gilbert. *A Sermon Preached at the Funeral of the Honourable Robert Boyle at St. Martins in the Fields, January 7, 1691/2.* London, 1692.

Burns, Norman T. *Christian Mortalism from Tyndale to Milton.* Cambridge: Harvard University Press, 1972.

Burns, R.M. *The Great Debate on Miracles: From Joseph Glanvill to David Hume.* Lewisburg, PA: Bucknell University Press, 1980.

Bynum, Caroline Walker. *The Resurrection of the Body in Western Christianity, 200–1336.* New York: Columbia University Press, 1994.

Calamy, Benjamin. *A Sermon Preached before the Right Honourable the Lord Mayor and the Court of Aldermen at Guild-Hall Chappel upon the 13th of July, 1673.* London, 1673.

Canny, Nicholas. *The Upstart Earl: A Study of the Social and Mental World of Richard Boyle, first Earl of Cork, 1566–1643.* Cambridge: Cambridge University Press, 1982.

Capp, B.S. *The Fifth Monarchy Men: A Study in Seventeenth-Century Millenarianism.* London: Faber, 1972.

Cassirer, Ernst. *The Platonic Renaissance in England,* translated by James P. Pettegrove. Austin: University of Texas Press, 1953.

Chadwick, Henry. *Augustine.* Past Masters. New York: Oxford University Press, 1986.

Chalmers, Alan. "The Lack of Excellency of Boyle's Mechanical Philosophy." *Studies in History of the Philosophy of Science* 24 (1993):541–564.

Champion, J.A.I. *The Pillars of Priestcraft Shaken: The Church of England and its Enemies 1660–1730.* Cambridge: Cambridge University Press, 1991.

Cheynell, Francis. *The Rise, Growth, and Danger of Socinianisme; together with a plaine discovery of a desperate design of corrupting the Protestant Religion, whereby it appeares that the Religion which hath been so violently contended for (by the Archbishop of Canterbury and his adherents) is not the true Protestant Religion, but an Hotchpotch of Arminianism, Socinianism and Popery: It is likewise made evident, that the Atheists, Anabaptists, and Sectaries so much complained of, have been raised or encouraged by the doctrines of the Arminian, Socinian and Popish Party.* London, 1643.

Chillingworth, William. *The Religion of Protestants a Safe Way to Salvation; or, An Answer to a Booke entituled Mercy and Truth; or, Charity Maintain'd by Catholiques, which Pretends to Prove the Contrary.* 10th edition, 3 vols. Oxford: 1838. Reprinted New York: AMS Press, 1972.

Clericuzio, Antonio. "Carneades and the Chemists: A Study of *The Sceptical Chymist* and its Impact on Seventeenth-Century Chemistry." In *Robert Boyle Reconsidered*, edited by Michael Hunter. Cambridge: Cambridge University Press, 1994.

——"From van Helmont to Boyle: A Study of the Transmission of Helmontian Chemical and Medical Theories in Seventeenth-Century England." *British Journal for the History of Science* 26 (1993):303–334.

——"A Redefinition of Boyle's Chemistry and Corpuscular Philosophy." *Annals of Science* 47 (1990):561–589.

Clifford, Alan C. *Atonement and Justification: English Evangelical Theology 1640–1790. An Evaluation.* Oxford: Clarendon Press, 1990.

Colie, Rosalie. *Light and Enlightenment: A Study of the Cambridge Platonists and the Dutch Arminians.* Cambridge: Cambridge University Press, 1957.

——"Spinoza in England 1665–1730." *Proceedings of the American Philosophical Society* 107 (1963):183–219.

Collier, Jeremy. *The Great Historical, Geographical, Genealogical and Poetical Dictionary.* 2nd edition. 2 vols. London, 1701.

Colligan, J. Hay. *The Arian Movement in England.* Manchester: University Press, 1913.

Cope, Jackson I. *Joseph Glanvill: Anglican Apologist.* St. Louis: Washington University Studies, 1956.

Copenhaver, Brian P. "Astrology and Magic," in *The Cambridge History of Renaissance Philosophy*, Charles B. Schmitt, general editor. Cambridge: Cambridge University Press, 1988.

——"Hermes Trismegistus, Proclus, and a Philosophy of Magic." In *Hermeticism and the Renaissance: Intellectual History and the Occult in Early Modern Europe*, edited by Ingrid Merkel and Allen G. Debus. Washington, DC: The Folger Shakespeare Library, 1988.

——and Charles B. Schmitt. *Renaissance Philosophy.* New York: Oxford University Press, 1992.

Courtenay, William J. "The Dialectic of Omnipotence in the High and Late Middle Ages." In *Divine Omniscience and Omnipotence in Medieval Philoso-*

phy, edited by Tamar Rudavsky. Dordrecht: D. Reidel Publishing Company, 1985.

Cragg, Gerald R. Introduction to *The Cambridge Platonists*, edited by Gerald R. Cragg. New York: Oxford University Press, 1968.

Craig, William Lane. *The Problem of Divine Foreknowledge and Future Contingents from Aristotle to Suarez*. Leiden: E.J. Brill, 1988.

Crellius, Johannes [Junius Brutus, pseudo.]. *Vindiciæ pro religionis libertate*, translated by John Dury under the title of *A Learned and exceedingly well-compiled Vindication of Liberty of Religion: Written by Junius Brutus in Latine, and Translated into English by N.Y. who desires, as much as in him is, to do good unto all men*. No Place [sic], 1646.

Cudworth, Ralph. *The True Intellectual System of the Universe: The First Part; Wherein, All the Reason and Philosophy of Atheism is Confuted; and its Impossibility Demonstrated*. Faksimile-Neudruck der Ausgabe von London 1678, Stuttgart-Badd Cannstatt: Friedrich Frommann Verlag (Günther Holzboog), 1964.

Daniel, Stephen H. *John Toland: His Methods, Manners, and Mind*. Kingston and Montreal: McGill-Queen's University Press, 1984.

D[anson], T[homas]. *De Causa Dei; or, A Vindication of the Common Doctrine of Protestant Divines, concerning Predetermination: (i.e., the Interest of God as the first Cause in all the Actions, as such, of all Rational Creatures:) from the invidious Consequences with which it is burdened by Mr. John Howe, in a late Letter and Postscript, of God's Prescience*. London, 1678.

Davis, Edward B. "The Anonymous Works of Robert Boyle and the *Reasons Why a Protestant Should not Turn Papist* (1687)." *Journal of the History of Ideas* 55 (1994):611–629.

——"Christianity and Early Modern Science: The Foster Thesis Reconsidered." In *The Evangelical Engagement with Science*, edited by Mark Noll, David Livingstone, and Daryl Hart. New York: Oxford University Press, forthcoming.

——"God, Man and Nature: The Problem of Creation in Cartesian Thought." *Scottish Journal of Theology* 44 (1991):325–348.

——"Newton's Rejection of the 'Newtonian World View': The Role of Divine Will in Newton's Natural Philosophy." *Fides et Historia* 22 (1990):6–20.

——"'Parcere nominibus': Boyle, Hooke and the Rhetorical Interpretation of Descartes." In *Robert Boyle Reconsidered*, edited by Michael Hunter. Cambridge: Cambridge University Press, 1994.

——and Michael P. Winship. "Early Modern Protestantism." In *The History of Science and Religion in the Western Tradition: An Encyclopedia*. Garland Publishing Co., forthcoming.

Deason, Gary. "Reformation Theology and the Mechanistic Conception of Nature." In *God and Nature: Historical Essays on the Encounter between Christianity and Science*, edited by David C Lindberg and Ronald L. Numbers. Berkeley and Los Angeles: University of California Press, 1986.

Debus, Allen G. "Fire Analysis and the Elements in the Sixteenth and the Seventeenth Centuries." *Annals of Science* 23 (1967):127–147. Reprinted (with same pagination) in Allen G. Debus. *Chemistry, Alchemy and the New Philosophy 1550–1700: Studies in the History of Science and Medicine.* London: Variorum Reprints, 1987.

Descartes, René. *Meditations on First Philosophy.* In *The Philosophical Works of Descartes,* translated by Elizabeth S. Haldane and G. R. T. Ross. 2 vols. Cambridge: Cambridge University Press, 1967, vol. 1.

Dobbs, Betty Jo Teeter. "Conceptual Problems in Newton's Early Chemistry: A Preliminary Study." In *Religion, Science, and Worldview: Essays in Honor of Richard S. Westfall,* edited by Margaret J. Osler and Paul Lawrence Farber. Cambridge: Cambridge University Press, 1985.

——*The Foundations of Newton's Alchemy, or "The Hunting of the Greene Lyon."* Cambridge: Cambridge University Press, 1975.

——*The Janus Faces of Genius: The Role of Alchemy in Newton's Thought.* Cambridge: Cambridge University Press, 1991.

Evelyn, John. "John Evelyn's Letter to William Wotton, 29 March 1696." In *Robert Boyle by Himself and His Friends,* edited by Michael Hunter. London: William Pickering, 1994.

Ferguson, Robert. *The Interest of Reason in Religion with the Import & Use of Scripture-Metaphors; and the Nature of the Union Betwixt Christ & Believers.* London, 1675.

Fisch, Harold. "The Scientist as Priest: A Note on Robert Boyle's Natural Philosophy." *Isis* 44 (1953):252–265.

Fisher, M.S. *Robert Boyle, Devout Naturalist: A Study in Science and Religion in the Seventeenth Century.* Philadelphia: Oshiver Studio Press, 1945.

Force, James E. "Newton's God of Dominion: The Unity of Newton's Theological, Scientific, and Political Thought." In *Essays on the Context, Nature, and Influence of Isaac Newton's Theology,* edited by James E. Force and Richard H. Popkin. Dordrecht: Kluwer Academic Publishers, 1990.

——*William Whiston: Honest Newtonian.* Cambridge: Cambridge University Press, 1985.

Foster, M.B. "The Christian Doctrine of Creation and the Rise of Modern Natural Science." *Mind* 43 (1934):446–468.

——"Christian Theology and Modern Science of Nature." Published in two parts in *Mind* 44 (1935):439–466 and *Mind* 45 (1936):1–27.

Fowler, Edward. *The Principles and Practices, Of certain Moderate Divines of the Church of England, (greatly mis-understood) Truly Represented and Defended: wherein . . . some controversies, of no mean importance, are succinctly discussed. . . .* London, 1670.

Frank, R.G. *Harvey and the Oxford Physiologists.* Berkeley and Los Angeles: University of California Press, 1980.

Gabbey, Alan. "Henry More and the Limits of Mechanism." In *Henry More (1614–1687): Tercentenary Studies,* edited by Sarah Hutton. Dordrecht: Kluwer Academic Publishers, 1990.

——"Philosophia Cartesiana Triumphata: Henry More (1646–1671)," in *Problems of Cartesianism,* edited by Thomas M. Lennon, John M. Nicholas, and

John W. Davis. Kingston and Montreal: McGill-Queen's University Press, 1982.

Gale, Theophilus. *The Court of the Gentiles, Part IV, Of Reformed Philosophie, wherein Plato's Moral, and Metaphysic, or prime Philosophie is reduced to an useful Forme and Method.* London, 1677.

Garber, Daniel. *Descartes' Metaphysical Physics.* Chicago: University of Chicago Press, 1992.

Gilson, Étienne. *The Christian Philosophy of St. Thomas Aquinas,* translated by I.T. Eschmann. New York: Random House, 1956.

——*History of Christian Philosophy in the Middle Ages.* New York: Random House, 1955.

Glanvill, Joseph. *Essays on Several Important Subjects in Philosophy and Religion.* London, 1676.

——*Lux Orientalis.* In *Two Choice and Useful Treatises: The One Lux Orientalis; Or, An Enquiry into the Opinion of the Eastern Sages Concerning the Præexistence of Souls, being a Key to unlock the Grand Mysteries of Providence. The Other, A Discourse of Truth,* annotated by Henry More with a preface by Joseph Glanvill. London, 1682.

ΛΟΓΟΥ ΘΡΗΣΚΕΙΑ; *or, A Seasonable Recommendation, and Defence of Reason, in the Affairs of Religion; Against Infidelity, Scepticism, and Fanaticisms of all sorts.* London, 1670.

Grant, Edward. "The Condemnation of 1277, God's Absolute Power, and Physical Thought in the Late Middle Ages." *Viator* 10 (1979):211–244.

——*Physical Science in the Middle Ages.* New York: John Wiley and Sons, 1971; paperback edition, Cambridge: Cambridge University Press, 1977.

——"Science and Theology in the Middle Ages." In *God and Nature: Historical Essays on the Encounter between Christianity and Science,* edited by David C. Lindberg and Ronald L. Numbers. Berkeley: University of California Press, 1986.

Grant, Robert M. "Tertullian." In the *Encyclopedia of Philosophy,* edited by Paul Edwards. 8 vols. New York: Macmillan and Free Press, 1969, vol. 8.

Greene, R.A. "Henry More and Robert Boyle on the Spirit of Nature." *Journal of the History of Ideas* 23 (1962):451–474.

Hacking, Ian. *The Emergence of Probability: A Philosophical Study of Early Ideas about Probability, Induction and Statistical Inference.* Cambridge: Cambridge University Press, 1975.

Hall, Marie Boas. "Boyle as a Theoretical Scientist." *Isis* 41 (1950):261–268.

——"The Establishment of the Mechanical Philosophy." *Osiris* 10 (1952):412–541.

——*Robert Boyle and Seventeenth-Century Chemistry.* Cambridge: Cambridge University Press, 1958.

——*Robert Boyle on Natural Philosophy: An Essay with Selections from his Writings.* Bloomington: Indiana University Press, 1965.

Hammond, Henry. ΧΑΡΙΣ ΚΑΙ ΕΙΡΗΝΗ; *or, A Pacifick discourse of Gods Grace and Decrees: In a Letter, of full Accordance written to the Reverend, and most learned, Dr. Robert Sanderson, to which are annexed the Extracts of three Letters concerning Gods Prescience reconciled with Liberty and Contingency.* London, 1660.

Hansen, Bert. "Science and Magic." In *Science in the Middle Ages,* edited by David C. Lindberg. Chicago: University of Chicago Press, 1978.

Harwood, John, editor. Introduction to *The Early Essays and Ethics of Robert Boyle.* Carbondale and Edwardsville: Southern Illinois University Press, 1991.

Heath, Thomas. *Mathematics in Aristotle.* Oxford: Clarendon Press, 1949.

Hefelbower, S. G. *The Relation of John Locke to English Deism.* Chicago: University of Chicago Press, 1918.

Henry, John. "Boyle and Cosmical Qualities." In *Robert Boyle Reconsidered,* edited by Michael Hunter. Cambridge: Cambridge University Press, 1994.

——"A Cambridge Platonist's Materialism: Henry More and the Concept of Soul." *Journal of the Warburg and Courtauld Institutes* 49 (1986):172–195.

——"Henry More versus Robert Boyle: The Spirit of Nature and the Nature of Providence." In *Henry More (1614–1687): Tercentenary Studies,* edited by Sarah Hutton. Dordrecht: Kluwer Academic Publishers, 1990.

——"Magic and Science in the Sixteenth and Seventeenth Centuries." In *Companion to the History of Modern Science,* edited by R.C. Olby, G.N. Cantor, J.R.R. Christie, and M.J.S. Hodge. London: Routledge, 1990.

——"Occult Qualities and the Experimental Philosophy: Active Principles in Pre-Newtonian Matter Theory." *History of Science* 24 (1986):335–381.

Heyd, Michael. "The Reaction to Enthusiasm in the Seventeenth Century: Towards an Integrative Approach." *Journal of Modern History* 53 (1981):258–280.

Hickman, Henry. *Historia Quinq-Articularis Exarticulata.* London, 1673.

Hill, Christopher. " 'Reason' and 'Reasonableness' in Seventeenth-Century England." *British Journal of Sociology* 20 (1969):235–252.

——*The World Turned Upside Down: Radical Ideas during the English Revolution.* New York: Viking Press, 1972.

Hoard, Samuel. *Gods Love to Mankind, manifested by disproving his absolute decree for their damnation.* Originally published 1633. London, 1673.

Hone, Richard B. *The Lives of Nicholas Ridley, D. D., Bishop of London; Joseph Hall, D. D., Bishop of Norwich; and The Honourable Robert Boyle.* Lives of Eminent Christians, 3. London, 1837.

Hoopes, Robert. *Right Reason in the English Renaissance.* Cambridge: Harvard University Press, 1962.

Hooykaas, R. *Religion and the Rise of Modern Science.* Edinburgh: Scottish Academic Press, 1972, first American edition, Grand Rapids, MI: William B. Eerdmans Publishing Co., 1972.

Hopkins, Jasper. *Nicholas of Cusa on Learned Ignorance: A Translation and an Appraisal of De Docta Ignorantia.* Minneapolis: Arthur J. Banning Press, 1981.

Hotchkiss, Thomas. *Reformation or Ruine: Being Certain Sermons upon Levit. XXVI.23, 24. First Preached, and afterwards with Necessary Enlargements fitted for Publick Use.* London, 1675.

Howe, John. *The Reconcileableness of God's Prescience of the Sins of Men, with the Wisdom and Sincerity of his Counsels, Exhortations, and whatsoever Means He uses to prevent them. In a Letter to the Honourable Robert Boyle Esq. To which is added a Postscript in Defence of the said Letter.* London, 1677.

Hume, David. *Enquiries Concerning Human Understanding and Concerning the Principles of Morals,* reprinted from the posthumous edition of 1777, edited by L.A. Selby-Bigge, 3rd edition, with text revised and notes by P.H. Nidditch. Oxford: Clarendon Press, 1975.

H[umphrey], J[ohn]. *The Middle-Way in One Paper of Election & Redemption, with Indifferency between the Arminian & Calvinist.* London, 1673.

Hunt, R. M. *The Place of Religion in the Science of Robert Boyle.* Philadelphia: University of Pittsburgh Press, 1955.

Hunter, Michael. "'Aikenhead the Atheist': The Context and Consequences of Articulate Irreligion in the Late Seventeenth Century." In *Atheism from the Reformation to the Enlightenment,* edited by Michael Hunter and David Wootton. Oxford: Clarendon Press, 1992.

——"Alchemy, Magic and Moralism in the Thought of Robert Boyle." *British Journal for the History of Science* 23 (1990):387–410.

——"Boyle versus the Galenists: A Suppressed Critique of Seventeenth-century Medical Practice and its Significance." *Medical History* 41 (1997).

——"Casuistry in Action: Robert Boyle's Confessional Interviews with Gilbert Burnet and Edward Stillingfleet, 1691." *Journal of Ecclesiastical History* 44 (1993):80–98.

——"The Conscience of Robert Boyle: Functionalism, 'Dysfunctionalism' and the Task of Historical Understanding." In *Renaissance and Revolution: Humanists, Scholars, Craftsmen and Natural Philosophers in Early Modern Europe,* edited by J.V. Field and F.A.J. James. Cambridge: Cambridge University Press, 1993.

——"How Boyle Became a Scientist." *History of Science* 33 (1995):59–103.

——"Latitudinarianism and the 'Ideology' of the Early Royal Society: Thomas Spratt's *History of the Royal Society* (1667) Reconsidered." In Michael Hunter, *Establishing the New Science: The Experience of the Early Royal Society.* Woodbridge, Suffolk: Boydell Press, 1989.

——*Letters and Papers of Robert Boyle: A Guide to the Manuscripts and Microfilm.* Bethesda, MD: University Publications of America, 1992.

——"The Reluctant Philanthropist: Robert Boyle and the 'Communication of Secrets and Receits in Physick'." In *Religio Medici,* edited by O. P. Grell and A. Cunningham. Aldershot, Hampshire: Scholar Press, 1996.

——*The Royal Society and its Fellows 1660–1700.* Chalfont St. Giles: British Society for the History of Science, 1982.

——"Science and Heterodoxy: An Early Modern Problem Reconsidered." In *Reappraisals of the Scientific Revolution,* edited by D.C. Lindberg and R.S. Westman. Cambridge: Cambridge University Press, 1990.

——*Science and Society in Restoration England.* Cambridge: Cambridge University Press, 1981; Aldershot, Hampshire: Gregg Revivals, 1992.

——Editor. *Robert Boyle by Himself and His Friends, with a fragment of William Wotton's lost "Life of Boyle."* London: William Pickering, 1994.

——Editor. *Robert Boyle Reconsidered.* Cambridge: Cambridge University Press, 1994.

——and Edward B. Davis. "The Making of Robert Boyle's *Free Enquiry into the Vulgarly Receiv'd Notion of Nature* (1686)." *Early Science and Medicine* 1 (1996):204–271.

——and Paul B. Wood. "Towards Solomon's House: Rival Strategies for Reforming the Early Royal Society." *History of Science* 24 (1986):49–108.

Hutchinson, Keith. "Supernaturalism and the Mechanical Philosophy." *History of Science* 21 (1983):297–333.

Hutton, Sarah Hutton. "Edward Stillingfleet, Henry More, and the Decline of Moses Atticus: A Note on Seventeenth-century Anglican Apologetics." In *Philosophy, Science, and Religion in England 1640–1700*, edited by Richard Kroll, Richard Ashcraft, and Perez Zagorin. Cambridge: Cambridge University Press, 1992.

——"Science, Philosophy, and Atheism: Edward Stillingfleet's Defence of Religion." In *Scepticism and Irreligion in the Seventeenth and Eighteenth Centuries*, edited by Richard H. Popkin and Arjo Vanderjagt. Leiden: E.J. Brill, 1993.

Irenaeus. *Against Heresies*, translated by Alexander Roberts and James Donaldson. In *The Ante-Nicene Fathers: Translations of the Writings of the Fathers down to A. D. 325*, edited by Alexander Roberts and James Donaldson, revised by A. Cleveland Cox. 10 volumes. American reprint of the Edinburgh Edition. Grand Rapids, MI: Eerdmans, 1957, vol. 1.

Jacob, James R. *Robert Boyle and the English Revolution: A Study in Social and Intellectual Change.* New York: Burt Franklin and Co., 1977.

——and Margaret C. Jacob. "The Anglican Origins of Modern Science: The Metaphysical Foundations of the Whig Constitution." *Isis* 71 (1980):251–267.

Kaplan, Barbara Beigun. *"Divulging of Useful Truths in Physick": The Medical Agenda of Robert Boyle.* Baltimore: The Johns Hopkins University Press, 1993.

Kargon, Robert Hugh. *Atomism in England from Hariot to Newton.* Oxford: Clarendon Press, 1966.

Klaaren, Eugene M. *Religious Origins of Modern Science.* Grand Rapids, MI: William B. Eerdmans Publishing Co., 1977.

Kneale, William, and Martha Kneale. *The Development of Logic.* New York: Oxford University Press, 1962.

Kroll, Richard. Introduction to *Philosophy, Science, and Religion in England 1640–1700.* Cambridge: Cambridge University Press, 1992.

Lake, Peter. "Anti-popery: the Structure of a Prejudice." In *Conflict in Early Stuart England: Studies in Religion and Politics 1603–1642*, edited by Richard Cust and Ann Hughes. London and New York: Longman, 1989.

Leff, Gordon. *Medieval Thought: St. Augustine to Ockham.* London: The Merlin Press, 1959.

Levine, Joseph M. *The Battle of the Books: History and Literature in the Augustan Age.* Ithaca, NY: Cornell University Press, 1991.

——"Latitudinarians, Neoplatonists, and the Ancient Wisdom." In *Philosophy, Science, and Religion in England 1640–1700,* edited by Richard Kroll, Richard Ashcraft, and Perez Zagorin. Cambridge: Cambridge University Press, 1992.

Loemker, Leroy M. "Boyle and Leibniz." *Journal of the History of Ideas* 16 (1955):22–43.

Lohr, Charles H. "Metaphysics." In *The Cambridge History of Renaissance Philosophy,* edited by Charles B. Schmitt, Quentin Skinner, Eckhard Kessler, and Jill Kraye. Cambridge: Cambridge University Press, 1988.

Maddison, R.E.W. *The Life of the Honourable Robert Boyle F.R.S.* London: Taylor and Francis, 1969.

Mandlebaum, Maurice. *Philosophy, Science and Sense Perception: Historical and Critical Studies.* Baltimore: The Johns Hopkins University Press, 1964.

Marshall, John. "The Ecclesiology of the Latitude-men 1660–1689: Stillingfleet, Tillotson and 'Hobbism'." *Journal of Ecclesiastical History* 36 (1985):407–427.

——*John Locke: Resistance, Religion and Responsibility.* Cambridge: Cambridge University Press, 1994.

——"John Locke and Latitudinarianism," in *Philosophy, Science, and Religion in England 1640–1700,* edited by Richard Kroll, Richard Ashcraft, and Perez Zagorin. Cambridge: Cambridge University Press, 1992.

[Marvell, Andrew]. *Remarks upon a Discourse writ by one T. D., under the pretence De Causa Dei, and of Answering Mr. John Howe's Letter and Postscript of God's Prescience, Affirming, as the Protestant's Doctrine, That God doth by Efficacious Influence universally move and determine Men to all their Actions, even to those that are most Wicked.* London, 1678.

Maurer, Armand. "Nicholas of Cusa." In *The Encyclopedia of Philosophy,* edited by Paul Edwards. 8 vols. New York: Macmillan and Free Press, 1969, vol. 5.

——Introduction to *St. Thomas Aquinas. Faith, Reason and Theology: Questions I–IV of his Commentary on the "De Trinitate" of Boethius.* Toronto: Pontifical Institute of Mediaeval Studies, 1987.

McGee, J. Sears. *The Godly Man in Stuart England: Anglicans, Puritans, and the Two Tables, 1620–1670.* New Haven, CT: Yale University Press, 1976.

McGregor, J.F., and B. Reay, editors. *Radical Religion in the English Revolution.* New York: Oxford University Press, 1984.

McGuire, J.E. "Atoms and the 'Analogy of Nature': Newton's Third Rule of Philosophizing." *Studies in History and Philosophy of Science* 1 (1970):3–58.

——"Boyle's Conception of Nature." *Journal of the History of Ideas* 33 (1972):523–542.

——"Force, Active Principles, and Newton's Invisible Realm." *Ambix* 15 (1968):154–208.

McInerny, Ralph. *Aquinas Against the Averroists: On There Being Only One Intellect.* West Lafayette, IN: Purdue University Press, 1993.

MacIntosh, J.J. "Locke and Boyle on Miracles and God's Existence." In *Robert Boyle Reconsidered,* edited by Michael Hunter. Cambridge: Cambridge University Press, 1994.

McLachlan, H. John. *Socinianism in Seventeenth Century England.* Oxford: Oxford University Press, 1951.

——"Studies in the Life of Robert Boyle, F.R.S.: Part VI, The Stalbridge Period, 1645–1655, and the Invisible College," *Notes and Records of the Royal Society of London,* 18 (1963):104–124.

——"Studies in the Life of Robert Boyle, F.R.S. Part VII. The Grand Tour." *Notes and Records of the Royal Society* 20 (1965):51–77.

Meinel, Christoph. "Early Seventeenth-Century Atomism: Theory, Epistemology, and the Insufficiency of Experiment." *Isis* 79 (1988):68–103.

Mendelson, Sara Heller. *The Mental World of Stuart Women: Three Studies.* Brighton, Sussex: Harvester Press, 1987.

Milton, Anthony. " 'The Unchanged Peacemaker'? John Dury and the Politics of Irenicism in England, 1628–1643." In *Samuel Hartlib and Universal Reformation: Studies in Intellectual Communication,* edited by Mark Greengrass, Michael Leslie, and Timothy Raylor. Cambridge: Cambridge University Press, 1994.

Molland, A. George. "Aristotelian Science." In *Companion to the History of Modern Science,* edited by R.C. Olby, G.N. Cantor, J.R.R. Christie, and M.J.S. Hodge. London: Routledge, 1990.

More, L.T. *The Life and Works of the Honourable Robert Boyle.* New York: Oxford University Press, 1944.

Morgan, Irvonwy. *The Nonconformity of Richard Baxter.* London: Epworth Press, 1946.

Morgan, John. *Godly Learning: Puritan Attitudes towards Reason, Learning, and Education, 1560–1640.* Cambridge: Cambridge University Press, 1986.

Mulligan, Lotte. "Robert Boyle, 'Right reason,' and the Meaning of Metaphor." *Journal of the History of Ideas* 55 (1994):235–257.

——" 'Reason', 'Right Reason', and 'Revelation'." In *Occult and Scientific Mentalities in the Renaissance,* edited by Brian Vickers. Cambridge: Cambridge University Press, 1984.

Muller, Richard A. *Christ and the Decree: Christology and Predestination in Reformed Theology from Calvin to Perkins.* Durham, NC: The Labyrinth Press, 1986.

Multhauf, Robert P. "The Science of Matter." In *Science in the Middle Ages,* edited by David C. Lindberg. Chicago: University of Chicago Press, 1978.

New, John F.H. *Anglican and Puritan: The Basis of their Opposition, 1558–1640.* Stanford, CA: Stanford University Press, 1964.

Newman, William R. "Boyle's Debt to Corpuscular Alchemy." In *Robert Boyle Reconsidered,* edited by Michael Hunter. Cambridge: Cambridge University Press, 1994.

——*Gehennical Fire: The Lives of George Starkey, an American Alchemist in the Scientific Revolution.* Cambridge: Harvard University Press, 1994.

Nicholas of Cusa on Interreligious Harmony: Text, Concordance and Transla-

tion of *De Pace Fidei*, edited by James E. Biechler and H. Lawrence Bond. Lewiston, NY: E. Mellen Press, 1990.

Norton, David Fate. "Hume and the Experimental Method." Unpublished typescript.

———"The Myth of British Empiricism." *History of European Ideas* 1 (1980):331–344.

Oakley, Francis. "Christian Theology and the Newtonian Science: The Rise of the Concept of Laws of Nature." *Church History* 30 (1961):433–457.

———*Omnipotence, Covenant, and Order: An Excursion in the History of Ideas from Abelard to Leibniz.* Ithaca, NY: Cornell University Press, 1984.

Ogonoswki, Zbigniew. "Le 'Christianisme sans mystères' selon John Toland et les sociniens." *Archiwum historii filozofii i mysli spolecznej* 12 (1966):pp. 205–223.

Orr, John. *English Deism: Its Roots and Fruits.* Grand Rapids, MI: Eerdmans, 1934.

Osborn, Eric. *The Emergence of Christian Theology.* Cambridge: Cambridge University Press, 1993.

Osler, Margaret J. "Baptizing Epicurean Atomism: Pierre Gassendi on the Immortality of the Soul." In *Religion, Science, and Worldview: Essays in Honor of Richard S. Westfall,* edited by Margaret J. Osler and Paul Lawrence Farber. Cambridge: Cambridge University Press, 1985.

———*Divine Will and the Mechanical Philosophy: Gassendi and Descartes on Contingency and Necessity in the Created World.* Cambridge: Cambridge University Press, 1994.

———"The Intellectual Sources of Robert Boyle's Philosophy of Nature: Gassendi's Voluntarism and Boyle's Physico-Theological Project." In *Philosophy, Science, and Religion in England, 1640–1700,* edited by Richard Kroll, Richard Ashcraft, and Perez Zagorin. Cambridge: Cambridge University Press, 1992.

Oster, Malcolm. "Biography, Culture, and Science: The Formative Years of Robert Boyle." *History of Science* 31 (1993):177–226.

———"Millenarianism and the New Science: The Case of Robert Boyle." In *Samuel Hartlib and Universal Reformation: Studies in Intellectual Communication,* edited by Mark Greengrass, Michael Leslie, and Timothy Raylor. Cambridge: Cambridge University Press, 1994.

———"Virtue, Providence and Political Neutralism: Boyle and Interregnum Politics." In *Robert Boyle Reconsidered,* edited by Michael Hunter. Cambridge: Cambridge University Press, 1994.

O'Toole, F.J. "Qualities and Powers in the Corpuscular Philosophy of Robert Boyle." *Journal of the History of Philosophy* 12 (1974):295–315.

Owen, John. *The Nature of Apostasie from the Profession of the Gospel, and the punishment of apostates declared, in an exposition of Heb. vi. 4–6, etc.* London, 1676.

———*Vindiciæ Evangelicæ; or, The Mystery of the gospell vindicated, and Socinianisme Examined, in the Consideration, and Confutation of a Catechisme, called A Scripture Catechisme, Written by J. Biddle M. A., and the Catechisme* of *Valentinus Smalcius, commonly called the Racovian Catechisme, with the Vindication of the Testimonies of Scripture, concerning the Deity and Satisfaction of Jesus Christ, from the Perverse Expositions, and Interpretations of them, by Hugo Grotius in his Annotations on the Bible; also an Appendix, in Vindication of some things formerly written about the Death of Christ, & the fruits thereof, from the Animadversions of Mr. R[ichard] B[axter].* London, 1655.

Passmore, John. *Ralph Cudworth: An Interpretation.* 1951; reprint Bristol: Thoemmes, 1990.

Patrides, C.A. Introduction to *The Cambridge Platonists,* edited by C.A. Patrides. Cambridge: Harvard University Press, 1970.

Pearl, Valerie. "London's Counter-Revolution." In *The Interregnum: The Quest for Settlement 1646–1660,* edited by G.E. Aylmer. London: Macmillan, 1972.

Pett, Sir Peter. *A Discourse Concerning Liberty of Conscience.* London, 1661.

——"Sir Peter Pett's Notes on Boyle." In *Robert Boyle by Himself and His Friends,* edited by Michael Hunter. London: William Pickering, 1994.

Plato. *The Euthyphro.* In *The Collected Dialogues of Plato,* edited by Edith Hamilton and Huntington Cairns. Princeton: Princeton University Press, 1961.

Pomponazzi, Pietro. *On the Immortality of the Soul,* translated by William Henry Hay II with revisions by John Herman Randall, Jr. In *The Renaissance Philosophy of Man,* edited by Ernst Cassirer, Paul Oskar Kristeller, and John Herman Randall, Jr. Chicago: University of Chicago Press, 1948.

Popkin, Richard H. "Hartlib, Dury and the Jews." In *Samuel Hartlib and Universal Reformation: Studies in Intellectual Communication,* edited by Mark Greengrass, Michael Leslie, and Timothy Raylor. Cambridge: Cambridge University Press, 1994.

——Introduction to Henry G. van Leeuwen, *The Problem of Certainty in English Thought, 1630–1690.* The Hague: Martinus Nijhoff, 1963.

——"Theories of Knowledge." In *The Cambridge History of Renaissance Philosophy,* edited by Charles B. Schmitt, Quentin Skinner, Eckhard Kessler, and Jill Kraye. Cambridge: Cambridge University Press, 1988.

——"The Philosophy of Bishop Stillingfleet." *History of Philosophy* 9 (1971):303–319.

——Introduction to *The Philosophy of the Sixteenth and Seventeenth Centuries,* edited by Richard H. Popkin. New York: Free Press, 1966.

——"The Third Force in 17th Century Philosophy: Scepticism, Science and Biblical Prophecy." *Nouvelles de la république des lettres* 3 (1983):35–63.

Poppi, Antonio. "Fate, Fortune, Providence and Divine Freedom," in *The Cambridge History of Renaissance Philosophy,* edited by Charles B. Schmitt, Quentin Skinner, Eckhard Kessler, and Jill Kraye. Cambridge: Cambridge University Press, 1988.

Principe, Lawrence M. *Aspiring Adept: Robert Boyle and his Alchemical Quest.* Princeton: Princeton University Press, forthcoming.

——"Boyle's Alchemical Pursuits." In *Robert Boyle Reconsidered,* edited by Michael Hunter. Cambridge: Cambridge University Press, 1994.

——"Boyle's Alchemical Secrecy: Codes, Ciphers and Concealments." *Ambix* 39 (1992):63–74.

——"Style and Thought of the Early Boyle: Discovery of the 1648 Manuscript of *Seraphic Love*." *Isis* 85 (1994):247–260.

Purver, Margery. *The Royal Society: Concept and Creation*. Cambridge: M.I.T. Press, 1967.

Ranger, Terence O. "Richard Boyle and the Making of an Irish Fortune." *Irish Historical Studies* 10 (1957):257–297.

Rattansi, P.M. "Paracelsus and the Puritan Revolution." *Ambix* 11 (1963):24–32.

Redondi, Pietro. *Galileo Heretic (Galileo Eretico)*. Princeton, NJ: Princeton University Press, 1987.

Redwood, John. *Reason, Ridicule and Religion: The Age of Enlightenment in England 1660–1750*. Cambridge: Harvard University Press, 1976.

Reedy, Gerard., S.J. *The Bible and Reason: Anglicans and Scripture in Late Seventeenth-Century England*. Philadelphia: University of Pennsylvania Press, 1985.

Reilly, Conor, S.J. *Francis Line S.J.: An Exiled English Scientist, 1595–1675*. Bibliotheca Instituti Historici S.I., 29. Rome: Institutum Historicum S.I., 1969.

Remarks on the Religious Sentiments of Learned and Eminent Laymen. London, 1792.

Rivers, Isabel. *Reason, Grace, and Sentiment: A Study in the Language of Religion and Ethics in England, 1660–1780*. Vol. 1, *Whichcote to Wesley*. Cambridge: Cambridge University Press, 1991.

Rogers, G.A.J. "Boyle, Locke, and Reason." *Journal of the History of Ideas* 27 (1966):205–216.

Rogers, Henry. *The Life and Character of John Howe, M.A., with an Analysis of his Writings*. London, 1836.

——Introduction to *Treatises on the High Veneration Man's Intellect owes to God; On Things Above Reason; and on The Style of the Holy Scriptures*, by Robert Boyle. The Sacred Classics, 18. London, 1835.

Rowbottom, M.E. "The Earliest Published Writing of Robert Boyle." *Annals of Science* 6 (1950):376–389.

Rust, George. *A Discourse of the Use of Reason in Matters of Religion: Shewing, That Christianity Contains nothing Repugnant to Right Reason; Against Enthusiasts and Deists*, annotated by Henry Hallywell. London, 1683.

Sargent, Rose-Mary. *The Diffident Naturalist: Robert Boyle and the Philosophy of Experiment*. Chicago: University of Chicago Press, 1995.

——"Learning from Experience: Boyle's Construction of an Experimental Philosophy." In *Robert Boyle Reconsidered*, edited by Michael Hunter. Cambridge: Cambridge University Press, 1994.

Schaff, Philip. *The Creeds of Christendom, with a History and Critical Notes*. 3 vols. New York: Harper, 1877.

Schmitt, Charles B. *Aristotle and the Renaissance*. Cambridge: Harvard University Press, 1983.

——"Reappraisals in Renaissance Science." In Charles B. Schmitt, *Studies in Renaissance Philosophy and Science*. London: Variorum Reprints, 1981.

Shanahan, Timothy. "God and Nature in the Thought of Robert Boyle." *Journal of the History of Philosophy* 26 (1988):547–569.
——"Teleological Reasoning in Boyle's *Disquisition about Final Causes.*" In *Robert Boyle Reconsidered,* edited by Michael Hunter. Cambridge: Cambridge University Press, 1994.
Shapin, Steven. *A Social History of Truth: Civility and Science in Seventeenth-Century England.* Chicago: University of Chicago Press, 1994.
——and Simon Schaffer. *Leviathan and the Air-Pump: Hobbes, Boyle, and the Experimental Life.* Princeton: Princeton University Press, 1985.
Shapiro, Barbara. *Probability and Certainty in Seventeenth-Century England.* Princeton: Princeton University Press, 1983.
Smith, Nigel. "The Charge of Atheism and the Language of Radical Speculation, 1640–1660." In *Atheism from the Reformation to the Enlightenment,* edited by Michael Hunter and David Wootton. Oxford: Clarendon Press, 1992.
Socinus, Faustus, et al. *The Racovian Catechisme: Wherein you have the substance of the Confession of those Churches, which in the Kingdom of Poland, and Great Dukedome of Lithuania, and other Provinces appertaining to that kingdom, do affirm, that no other save the Father of our Lord Jesus Christ, is that one God of Israel, and that the man Jesus of Nazareth, who was born of the Virgin, and no other besides, or before him, is the onely begotten Sonne of God,* translated by John Biddle. Amsterledam [London], 1652.
Spurr, John. " 'Rational Religion' in Restoration England." *Journal of the History of Ideas* 49 (1988):563–585.
——*The Restoration Church of England, 1646–1689.* New Haven, CT: Yale University Press, 1991.
Stegmann, Joachim. *Brevis Disquisitio: Or, a Brief Enquiry Touching a Better way Then is commonly made use of, to refute Papists, and reduce Protestants to certainty and Unity in Religion,* translated by John Biddle. London, 1653.
Stephen, Leslie. *History of English Thought in the Eighteenth Century.* 3rd edition. 2 vol. New York: G.P. Putnam's Sons, 1902, vol. 1.
Stewart, Larry R. *The Rise of Public Science: Rhetoric, Technology, and Natural Philosophy in Newtonian Britain, 1660–1750.* Cambridge: Cambridge University Press, 1992.
Stewart, M.A. Introduction to *Selected Philosophical Papers of Robert Boyle,* M.A. Stewart, editor. Originally published by Manchester University Press, 1979; Indianapolis, IN: Hackett Publishing Co., 1991.
Stillingfleet, Edward. *Origines Sacræ; or, A Rational Account of the Grounds of Christian Faith, as to the Truth and Divine Authority of the Scriptures, and the matters therein contained.* London, 1662.
Sullivan, Robert E. *John Toland and the Deist Controversy.* Cambridge: Cambridge University Press, 1982.
Taylor, Jeremy. *Ductor Dubitantium or the Rule of Conscience in all her generall measures.* 2 vols. London, 1660.
Tazbir, Janusz. "Poland." In *The Reformation in National Context,* edited by Bob Scribner, Roy Porter, and Mikulás Teich. Cambridge: Cambridge University Press, 1994.

Tertullian. *On the Flesh of Christ,* translated by Peter Holmes. In *The Ante-Nicene Fathers: Translations of the Writings of the Fathers down to A.D. 325,* edited by Alexander Roberts and James Donaldson, revised by A. Cleveland Cox. 10 volumes. American reprint of the Edinburgh Edition. Grand Rapids, MI: Eerdmans, 1957, vol. 3.

Thomas, Keith. *Religion and the Decline of Magic: Studies in Popular Beliefs in Sixteenth and Seventeenth Century England.* London: Weidenfeld and Nicolson, 1971.

Toland, John. *Christianity not Mysterious; or, A Treatise shewing that there is nothing in the Gospel contrary to reason, nor above it, and that no Christian doctrine can be properly call'd a mystery.* London, 1696.

Toon, Peter. *God's Statesman: The Life and Works of John Owen, Pastor, Educator, Theologian.* Exeter: Paternoster Press, 1971.

Trevor-Roper, Hugh. *Catholics, Anglicans and Puritans: Seventeenth Century Essays.* London: Secker and Warburg, 1987.

Tumbleson, Raymond D. " 'Reason and Religion': The Science of Anglicanism." *Journal of the History of Ideas* 57 (1996):131–156.

Tyacke, Nicholas. *Anti-Calvinists: The Rise of English Arminianism c. 1590–1640.* Oxford: Clarendon Press, 1987.

Valla, Lorenzo. *Dialogue on Free Will,* translated by Charles Edward Trkinkaus, Jr. In *The Renaissance Philosophy of Man,* edited by Ernst Cassirer, Paul Oskar Kristeller, and John Herman Randall, Jr. Chicago: University of Chicago Press, 1948.

van Leeuwen, Henry G. *The Problem of Certainty in English Thought, 1630–1690.* The Hague: Martinus Nijhoff, 1963.

Wallace, Dewey D., Jr. *Puritans and Predestination: Grace in English Protestant Theology, 1525–1695.* Chapel Hill: University of North Carolina Press, 1982.

Walton, Izaak. *Lives of John Donne, Henry Wotton, Richard Hooker, and George Hebert.* London: George Routledge and Sons, 1888.

Watson, Richard A. "Transubstantiation among the Cartesians." In *Problems of Cartesianism,* edited by Thomas M. Lennon, John M. Nicholas, and John W. Davis. Kingston and Montreal: McGill-Queen's University Press, 1982.

Webster, Charles. *The Great Instauration: Science, Medicine and Reform, 1626–1660.* London: Duckworth, 1975.

Westfall, Richard S. *Never at Rest: A Biography of Isaac Newton.* Cambridge: Cambridge University Press, 1980; first paperback edition, 1983.

——*Science and Religion in Seventeenth-Century England.* New Haven, CT: Yale University Press, 1958.

White, Peter. *Predestination, Policy and Polemic: Conflict and Consensus in the English Church from the Reformation to the Civil War.* Cambridge: Cambridge University Press, 1992.

Williams, George Huntston. "Michael Servetus." In *The Encyclopedia of Philosophy,* edited by Paul Edwards. 8 vols. New York: Macmillan and Free Press, 1967, reprint edition, 1972, vol. 7.

——*The Radical Reformation.* Philadelphia: Westminster Press, 1962.

——Editor and translator. *The Polish Brethren: Documentation of the History and Thought of Unitarianism in the Polish-Lithuanian Commonwealth and in the Diaspora, 1601–1685.* 2 vols. Missoula, MT: Scholars Press, 1980.

Wilson, Catherine. *The Invisible World: Early Modern Philosophy and the Invention of the Microscope.* Princeton: Princeton University Press, 1995.

Winter, Ernst F., editor and translator. *Erasmus-Luther Discourse on Free Will.* New York: Frederick Ungar Publishing Co., 1974.

Wojcik, Jan W. "'Behold the Fear of the Lord': The Erastianism of Stillingfleet, Wolseley, and Tillotson." In *Heterodoxy, Spinozism and Freethought,* edited by S. Berti, F. Charles-Daubert, and R. H. Popkin. Dordrecht: Kluwer Academic Publishers, 1996.

——"Pursuing Knowledge: Robert Boyle and Isaac Newton." In *The Canonical Imperative: Rethinking the Scientific Revolution,* edited by Margaret J. Osler, forthcoming.

——"The Theological Context of Boyle's *Things above Reason.*" In *Robert Boyle Reconsidered,* edited by Michael Hunter. Cambridge: Cambridge University Press, 1994.

——"Review of *Philosophy, Religion, and Science, 1640–1700,*" edited by Richard Kroll, Richard Ashcraft, and Perez Zagorin. Cambridge: Cambridge University Press, 1992. In *Journal of the History of Philosophy* 32 (1994):141–142.

Wybrow, Cameron, editor. *Creation, Nature, and Political Order in the Philosophy of Michael Foster (1903–1959): The Classic "Mind" Articles and Others, with Modern Critical Essays.* Lewiston: Edwin Mellon, 1992.

Yates, Frances A. *Giordano Bruno and the Hermetic Tradition.* Chicago: University of Chicago Press, 1964.

Index

Generally, where text discussions have been continued or amplified in the footnotes, separate entries for the notes have not been included.

Act of Uniformity (1662), 21
Adam, fall of 37, 80; Anglicans on, 39; Boyle on, 104; Baxter on, 65; Ferguson on, 69, 70, 71; Puritans on, 38; Socinians on, 45, 88
Adamites, 17n
afterlife, 10, 210–211
air, spring and weight of, 165, 166, 173, 175, 176–178. *See also* Boyle's Law
air-pump, 173, 174
alchemy, Boyle and, 122, 123, 129, 133–135; and corpuscularianism, 133–134; matter theory of, 131–132; as sacred knowledge, 135; theological concerns of, 132. *See also* tria prima
Alexandria, Clement of, 29, 36
Alsted, Johann, 203
Anabaptists, 14
angels, activity of, 138, 143, 174, 182; existence of, 140, 141, 142–143; intellect of, 143–144, 189, 207, 209
Anglicans, and reason's competence, 39. *See also* Church of England; latitudinarianism
antinomianism, 10n, 81, 83n, 113
Aquinas, Thomas, on double-truth, 31–32; on reason and religion, 29–31, 36, 97, 200; on reason's competence, 39, 40; on the soul, 125; on the Trinity, 31
Arianism, 45n
Aristotelianism, matter theory of, 123–124, 131, 159, 182; theological problems of, 29–30, 32n, 33n, 34, 97, 122–123, 124–126, 200. *See also* substantial forms
Aristotelians, Christian concerns of, 122–123; on creation of the world, 196n; on torrid zone, 161
Aristotle, on eternity of the world, 139; on the heavens, 161; modest claims of,

186; Tertullian on, 27; on unity of the intellect, 32n
Arminianism; Boyle on, 83–84; Howe's "middle-way," 23, 75; as implying contradictions, 82, 160; and predestination controversies, 8, 14, 24, 91, 213; and Socinianism, 48
Arminians, Dutch, 47
Athanasians, 45n
atheism, Boyle on, 20, 24n, 97–98, 122, 142–143, 153–154, 185; Cudworth on, 127; Glanvill on, 142–143
Augustine, 49, 154
Averroës, 200
Averroists, 30, 31, 32n, 33n

Bacon, Francis, 164; "Idols of the Tribe," 98
Barlow, Thomas, 17, 91
Baxter, Richard, 59, 62–65, 67, 68, 92, 100, 114, 195n
Bayle, Pierre, 52n
Beale, John, 22–23, 48n
Beck, Daniel A., 209n
Bentley, Richard, 2n
Berkeley, George, x
Biddle, John, 50–55, 59–60, 60, 61
Birch, Thomas, 10–11, 23n, 212
Boas, Marie. *See* Hall, Marie Boas
Boyle, Charles, 2n
Boyle, Lewis (Viscount Kinalmeaky), 12
Boyle, Richard (1st Earl of Burlington), 12
Boyle, Richard (1st Earl of Cork), 12, 201
Boyle, Robert, difficulty of classifying works, 2, 117n, 151; early ethical writings, 2n; funeral sermon of, 1, 10–11, 18n; gentlemanliness of, 24n, 58; and irenicism, 10–11, 19–24, 59n, 101, 122, 123n; library of, 115n; on openness of communication, 134–135; philological

Boyle, Robert (*cont.*)
  studies of, 57n, 84, 95, 97; and "Reasons Why a Protestant Should not Turn Papist," 78n; 106
Boyle, Roger (Baron Broghill), 13, 21n, 22
Boyle's Law, 177. *See also* air, spring and weight of
Budgell, Eustace, 1
Burnet, Gilbert, 1, 10–11, 18n, 212
Burnet, Thomas, 140n

Calamy, Benjamin, 39n, 115–116
Calvin, John, 44, 201, 202
Calvinism, Boyle on, 83–84, 108, 113; Hammond on, 91–92; Howe on, 23, 91–92; as implying contradictions, 82, 160. *See also* entries under individual Calvinists
Cambridge Platonists, Boyle on, 122–123; and original sin, 39n; rationalism of, 39
Cartesianism, 70, 206
causes, scale of, 156, 165, 166, 168, 177
Charles I, 13
Charles II, 13
Cheynell, Francis, 48–49
Chillingworth, William, 42–43
Christianity, mysteries of, 28, 36–37; Boyle on, 96–97, 100, 108, 117, 189; Ferguson on, 69–72, 74–75; Glanvill on, 67–68; Stillingfleet on, 174
chrysopoeia. See transmutation
Church of England, 14, 17, 21, 72, 76; Boyle and, 215, 216; Thirty-Nine Articles of, 68; *See also* Anglicans, latitudinarianism
Civil War, 17, 59, 85; Boyle and, 11–14; and illuminist epistemology, 37n
Clifford, Alan C., 76n
Collier, Jeremy, 1
Comenius, Jan, 202–203
conscience, liberty of, 11, 15, 50, 217. *See also* toleration
Condemnations (of 1270, 1277), 32–33, 200
contradictions, scriptural, Baxter on, 63–65; Boyle on, 98, 103, 112–114; Ferguson on, 69, 70–71; Glanvill on, 67–68; Owen on, 72–73
corpuscular hypothesis, and alchemical matter theory, 133–134; based on phenomena, 175; and Boyle's theology, 122; insufficiency of, 184, 185, 186; truth of, 172; usefulness of, 187
Creeds (of Christianity), 68
Crellius, Johannes, 16, 46, 50, 57, 60
criterion, problem of, 52–53

Cromwell, Oliver, 59
Cromwell, Richard, 13, 17, 89
Cudworth, Ralph, 127–129, 183n, 199n

Damian, Peter, 32n
Danson, Thomas, 93–94
Davis, Edward B., 191n, 204, 206n
deism, 6–7, 47
Descartes, René, 125n, 146, 153, 180n, 182–183, 197–200, 205
dialogue, Boyle's use of, 11, 101, 213
Diodati, Jean, 13n, 202
dogmatism, Boyle on, 19, 24n
double-truth, 31–33, Boyle on, 98–99
Dudithius, Andreas, 23n, 56nn
Dury, John, 15, 16, 50, 55, 202–203

empiricism, x, 3, 146–147
enthusiasm, 8, 115, 192n
Epicureanism, 97–98, 140, 149, 159, 182, 196n, 206, 208n
Epicurus, 186
Evelyn, John, 18n
evil, problem of, Boyle on, 128n; Cudworth on, 127–128

fact, matter of, 161, 165, 178n, 218
Familiasts, 14
Ferguson, Robert, 59, 69–72, 74–75, 92, 98, 100, 103, 108, 109, 114, 115
final causes, 4, 146, 206
Fisher, M.S., 105, 214
Force, James E., 92n
"Foster thesis," 190. *See also* voluntarism
free will, 48–49, 81, 82, 85, 89–90, 92

Gabbey, Alan, 184
Gale, Theophilus, 93
Galileo, 101, 138n, 153
Gassendi, Pierre, 97, 154n, 199, 203
Gilson, Étienne, 27, 151
Glanvill, Joseph, 6, 7, 59, 65–68, 69, 71, 73, 86, 92, 108, 142–143, 194–196, 203, 216
Gnosticism, 28, 29, 130
God, assists natural philosophers, 142; attributes of, 84n, 102–103, 190, 194, 200–201, 204–206; as author of sin, 81, 82, 88, 92–93, 93–94; and decrees of predestination, 35; exhortations of, 81, 82; existence of, 4, 149; and free will of humans, 35, 40, 102, 107–108, 113, 115; goodness and mercy of, 81, 82; grace of, 85, 91–92; immateriality of, 111; omnipotence of, 5, 30, 32, 33, 35, 58n, 81, 82, 98–100; 102–103;

God (*cont.*)
162–163, 171–172, 179; prescience of, 35, 40, 80, 82, 85, 91, 92, 102, 107–108, 110, 112, 113, 115; two books of, 3; wisdom of, 82
Grant, Robert M., 27n
gravity, 173
Great Tew Circle, 43n

Hall, Marie Boas, 213
Hallywell, Henry, 108, 109n, 192n, 193n
Hammond, Henry, 90–92, 94, 114
Hartlib, Samuel, 16n, 202–203
Hartlib Circle, 12, 13n
Harwood, John, xivn, 115n
Hefelbower, S.G., 214
Helmont, Jan Baptista van, 132
Helmontians, 132, 133n, 142
Hermeticism, 130
historiography, and contextual studies, xi–xii, 136, 213; Whiggish, ix
Hoard, Samuel, 90
Hobbes, Thomas, 89–90, 111, 113, 122, 140, 162–163, 178n, 218
Hone, Richard B., 2
Hooykaas, R., 198
Hotchkiss, Thomas, 90
Hottinger, J.C., 55n
Howe, John, and Boyle, 75, 86–87; nonconformity of, 85–86; Calvinist responses to, 8, 23, 24, 93–94, 100
Hume, David, x, 5n
Humphrey, John, 79–80, 89
Hunt, R.M., 214
Hunter, Michael, 17n, 143
hylarchic principle, 126, 128, 166, 172–174, 175, 178–179, 179. *See also* plastick nature
hylomorphism. *See* Aristotelianism, matter theory of
hypotheses, consistency of, 174–175, 176n; falsity of, 9 168, 175–179, 189; funicular, 176–177; good and excellent, 9, 166–167, 168, 176, 179; intelligibility of, 168–171; and matters of fact 218; as natural explanations, 162–163; provisional nature of, 164–165, 165–166; simplicity of, 171–174, 175, 181; sufficient evidence for, 163–166, 175; truth of, 168, 172, 189. *See also* corpuscular hypothesis

ideas, innate, 190, 198
imperceptibles, 147–150, 157, 179–180
Independents, 14, 17, 59
infinites, 153–154

interpretation, scriptural, 42, 64, 76–77, 79, 80n; Boyle on, 8, 14, 19, 56–58, 83, 110–112, 116; Calvinist, 7–8, 38–39, 109, 115
Interregnum. *See* Civil War
Irenaeus, 28–29, 40, 151
Islam religion, 34
Israel, Menassah ben, 84

Jacob, James R., 17n, 18n, 59n, 217, 218
Jacob, Margaret C., 217, 218
James I, 47
Jews, 57, 84, 85

Klaaren, Eugene M., 204
knowledge, construction of, 218; demonstrative, 30; a priori, 33

latitudinarianism, 62, 73–74, 215–216, 217. *See also* Anglicans, Church of England
Laud, William, 48, 49, 50n
Line, Francis, 137–138, 162, 163, 176–177
Locke, John, 6, 180n
logic, principles of, 103, 104, 114. *See also* noncontradiction, law of
Luther, Martin, 76, 201, 203
Lutherans, 16

McGuire, J.E., 204
macrocosm/microcosm, 170
Marcombes, Isaac, 13, 14, 17, 202
Marvell, Andrew, 93–94
mathematics, laws of, 197
matter, infinite divisibility of, 159–160, 183
Melancthon, Philipp, 203
microscope, 147, 172
millenarianism, 202–203, 211n
miracles, 5, 58, 66, 191, 203–204
Montaigne, Michel, 52n, 151–152
moral certainty, 217
More, Henry, 39n, 126–127, 128, 140n, 143n, 172–174, 178–179, 192n, 195, 203
More, L.T., 106
Moskorzowski, Jeromos, 45n
Mulligan, Lotte, 37–38n, 216

nature, a priori knowledge of, 198; laws of, 196, 198; mysteries of, 28, 67, 96, 117, 153–154, 156–157, 161, 184–186, 188, 189, 210, 212, 219
Neoplatonism, 130
Newton, Isaac, x, 138n, 201

Nicholas of Cusa, 34, 37, 40
nominalism, 33–34, 37, 40, 200
nonconformists, 7–8, 18n, 39–40, 74–75, 215
noncontradiction, law of, 32–34, 82; Boyle on, 8, 34, 39–40, 100, 102–103, 103–107, 107–108, 146

Ockhamists. *See* nominalism
Ockham's razor. *See* hypotheses, simplicity of
Ockham, William of, 33–34, 200
Osler, Margaret J., 191n, 198, 204
Owen, John, 38, 59–62, 69, 71n, 72–73, 74–75, 92, 100, 109, 114, 115

pantheism, 111n
Paracelsianism, 130, 131–132, 133, 134, 136, 142, 179. *See also* alchemy
Pascal, Blaise, 52n, 199
Pelagianism, 48–49, 80, 81
Perkins, William, 202
Pett, Peter, 16n, 17, 50n
plastick nature, 127–128, 189. *See also* hylarchic principle
Plato, 192n
Polish Brethren. *See* Socinians
Pomponazzi, Pietro, 33n
potentia Dei, absoluta and *ordinata*, 191–192
predestination, absolute, 79, 80–82, 91, 92–93, 93–94, 160; conditional, 79, 80–82, 87–88, 91, 160
Presbyterians, 14, 17
Principe, Lawrence M., 84–85, 129, 135
probabilism, 215–216
prophecies, 5n, 58n, 85
Protestants, 77, 80. *See also* particular denominations
psychopannychism, 44
Puritans, 38–39, 130

Quakers, 17n
qualities, primary and secondary, 178n

Racovian Catechism, 45, 47–48, 49, 53–54, 57, 60, 88, 93
Ranelagh, Katherine, 12, 13n, 16n
rarefaction, 176–177
reason, abstract and concrete, 69, 98; disparity between God's and man's, 104, 205–206, 208, 209; God as author of, 206–209
Rebellion, Irish, 12
Reformation, 42–43, 61, 132. *See also* Socinianism

Remonstrants, 85. *See also* Arminianism
Restoration, 17, 21, 38
resurrection, 20, 22n, 56, 58n, 99, 125, 140, 141
Rich, Mary (Countess of Warwick), 12
"right reason," 37–41; Boyle on, 97, 100, 101–102, 216
Rogers, Henry, 2, 11
Roman Catholicism, 16, 17, 21, 44n, 42–43, 46, 48–49, 50, 52–53, 56, 57, 72, 77, 81, 84, 85, 124–125
rota Aristotelica, 138n
Royal Society, 18n, 152, 164n, 218
rule of faith controversy, 43n
Rushd, Ibn. *See* Averroës
Rust, George, 192–194, 195–196

Sabellianism, 45n, 48
Sanderson, Robert, 90–91
Sargent, Rose-Mary, 204n, 218n, 219n
Satan, 61
Schaffer, Simon, 101n, 146–147, 165, 178n, 217–218
Schlichting, Jonas, 46, 60
scripture, as allegory, 140; duty to study, 56, 57; passages cited, 1 Corinthians 3:2, 29; 1 Corinthians 15:21, 54n; Genesis 2:19, 54; Genesis 31:8–11, 142n; James 2:24, 79; James 3:6, 140n; John 1:1, 61; John 1:14, 54, 61; John 3:9, 62; Leviticus 26:23–24, 90; Matthew 26:26, 43; 1 Peter 2:2, 51; 2 Peter 3:7, 10, 13, 141n; Philippians 4:6, 55; Romans 11:33, 35; Romans 12:1, 51; Romans 3:28, 79; 1 Timothy 2:5, 54n; 1 Timothy 5:21, 79; as rule of faith, 42, 43n, 46, 52–53; in the vernacular, 21, 56. *See also* contradictions, scriptural; interpretation, scriptural
sects, 15, 17, 18n, 42, 98
Servetus, Michael, 44
Shanahan, Timothy, 204
Shapin, Steven, 101n, 146–147, 165, 178n, 214n, 217–218
Shapiro, Barbara, 215–216
Sherlock, William, 69, 71, 72
Siger of Brabant, 31
Smalcius, Valentine, 45n, 60
Socinians, 17n, 84, 85, 108, 112
Socinianism, 115, 213; anti-Trinitarianism of, 45n, 53–54, 58, 60; Boyle on, 19–23; and enthusiasm, 48; and free will, 47, 48; and predestination, 46, 54–55, 87, 88–89; and prescience of God, 112n, 113; and scriptural interpretation, 6–7, 46, 57; rapid spread in England, 7,

Socinianism (*cont.*)
40–41; and Radical Reformation, 43–44; rationalism of, 8, 14, 24, 40, 42, 43, 44–46, 51–55, 85, 101; and religious liberty, 46–47; and resurrection, 55n, 141n
Socinus, Faustus, 23n, 44–45, 56nn, 57, 61
Socinus, Laelius, 44
soul, immortality of 4, 125, 141n, 172. *See also* resurrection
Spinoza, 90, 111, 113, 122, 213
spirit-contact, 3, 135, 141–142, 142–143
Starkey, George, 142n
Statorius, Peter Jr., 45n
Stegmann, Joachim, 87–88
Sterry, Peter, 89
Stillingfleet, Edward, 73–74, 87, 98, 103, 140n
Stoicism, 130, 140
sublapsarianism, 80
substantial forms, 124, 125, 126, 128, 166, 170–171, 172, 175, 179, 181, 182, 187, 189
supralapsarianism, 80, 202n

Tallents, Francis, 20
Taylor, Jeremy, 38n
Tempier, Étienne, 32–33
Tertullian, 27, 40
Test Acts, 18n
theology, mysteries of, 63, 156–157, 161, 169, 210, 212, 216, 219; natural, 4–6, 40, 213, 217
Toland, John, 151n
toleration, 15, 17, 18–19, 46–47. *See also* conscience, liberty of
Toricellian experiment, 162
transmutation, 129, 133, 135
Transubstantiation, 42–43, 51, 124–125
tria prima, 131, 133, 134, 179, 182
Trinity, 67, 68, 70, 156
Tumbleson, Raymond D., 78n

Union, of Christ with believers, 69, 71–72, 74–75
Ussher, James, 55, 97

vacuum, 124, 170, 174–175, 178, 181, 183
Valla, Lorenzo, 35–36, 37, 40, 92n
void. *See* vacuum
Volkelius, Johannes, 45n, 57, 60
voluntarism, and Reformation theology, 200–201

Warr, John Jr., 115n
Webster, John, 87n
Westfall, Richard S., 209n, 214
William of Ockham. *See* Ockham, William of
Worsley, Benjamin, 55
Wotton, Henry, 16

Printed in the United States
By Bookmasters